WHAT IS THE NAME OF THIS BOOK?

The Riddle of Dracula and Other Logical Puzzles

RAYMOND M. SMULLYAN

DOVER PUBLICATIONS, INC.
MINEOLA, NEW YORK

Bibliographical Note

This Dover edition, first published in 2011, is an unabridged republication of the work originally published by Prentice-Hall, Inc., Englewood Cliffs, N.J., in 1978.

Library of Congress Cataloging-in-Publication Data

Smullyan, Raymond M.
    What is the name of this book? : the riddle of Dracula and other logical puzzles / Raymond M. Smullyan.
        p. cm.
    Orig. pub.: New York : Simon & Schuster, c1978.
    Summary: "In his most critically acclaimed work, a well-known mathematician, magician, and author spins a logical labyrinth of more than 200 increasingly complex and challenging problems - puzzles that delve into some of the deepest paradoxes of logic and set theory. Solutions. "The most original, most profound, and most humorous collection of recreational logic and math problems ever written." – Martin Gardner"— Provided by publisher.
    ISBN-13: 978-0-486-48198-2 (pbk.)
    ISBN-10: 0-486-48198-0 (pbk.)
    1. Logic puzzles. I. Title.

GV1493.S63 2011
793.73—dc22

2011006160

Manufactured in the United States by LSC Communications
48198008   2017
www.doverpublications.com

Dedicated to
Linda Wetzel and Joseph Bevando,
whose wise counsels have been invaluable.

Contents

Part Four • LOGIC IS A MANY-SPLENDORED THING

My Thanks to

First I wish to thank my friends Robert and Ilse Cowen and their ten-year-old-daughter, Lenore, who went through this manuscript together and provided many helpful suggestions. (Lenore, incidentally, suspected all along the true answer to the key question of Chapter 4: Does Tweedledoo really exist, or is he merely a fabrication of Humpty Dumpty?)

I am grateful to Greer and Melvin Fitting (authors of the charming and useful book *In Praise of Simple Things*) for their kindly interest in my work and for having called it to the attention of Oscar Collier of Prentice-Hall. I also think Melvin should be thanked for actually appearing in this book (thereby refuting my proof that he couldn't appear!).

It was a pleasure working with Oscar Collier and others at Prentice-Hall. Mrs. Ilene McGrath who first copy-edited the text made many suggestions which I have gratefully adopted. I thank Dorothy Lachmann for her expert handling of production details.

I wish to again mention my two dedicatees, Joseph Bevando and Linda Wetzel, who have been heart and soul with this book from its very inception.

My dear wife, Blanche, has helped me with many a query. It is my hope that this book will enable her to decide whether she is married to a knight or a knave.

PART ONE
Logical Recreations

1 ○ Fooled?

1. Was I Fooled?

My introduction to logic was at the age of six. It happened this way: On April 1, 1925, I was sick in bed with grippe, or flu, or something. In the morning my brother Emile (ten years my senior) came into my bedroom and said: "Well, Raymond, today is April Fool's Day, and I will fool you as you have never been fooled before!" I waited all day long for him to fool me, but he didn't. Late that night, my mother asked me, "Why don't you go to sleep?" I replied, "I'm waiting for Emile to fool me." My mother turned to Emile and said, "Emile, will you please fool the child!" Emile then turned to me, and the following dialogue ensued:

> *Emile* / So, you expected me to fool you, didn't you?
> *Raymond* / Yes.
> *Emile* / But I didn't, did I?
> *Raymond* / No.
> *Emile* / But you expected me to, didn't you?
> *Raymond* / Yes.
> *Emile* / So I fooled you, didn't I!

Well, I recall lying in bed long after the lights were turned out wondering whether or not I had really been fooled. On the one hand, if I wasn't fooled, then I did not get what I expected, hence I was fooled. (This was Emile's argument.) But with equal reason it can be said that if I was fooled, then I *did* get what I expected, so then, in what sense was I fooled. So, was I fooled or wasn't I?

I shall not answer this puzzle now; we shall return to it in one form or another several times in the course of this book. It embodies a subtle principle which shall be one of our major themes.

2. Was I Lying?

A related incident occurred many years later when I was a graduate student at the University of Chicago. I was a professional magician at the time, but my magic business was slow for a brief period and I had to supplement my income somehow. I decided to try getting a job as a salesman. I applied to a vacuum cleaner company and had to take an aptitude test. One of the questions was, "Do you object to telling a little lie every now and again?" Now, at the time I definitely *did* object—I particularly object to salesmen lying and misrepresenting their products. But I thought to myself that if I truthfully voiced my objection, then I wouldn't get the job. Hence I lied and said "No."

Riding back home after the interview, I had the following thoughts. I asked myself whether I objected to the lie I had given to the sales company. My answer was "No." Well, now, since I didn't object to that particular lie, then it follows that I *don't* object to all lies, hence my "No" answer on the test was not a lie, but the truth!

To this day it is not quite clear to me whether I was lying or not. I guess logic might require me to say that I was telling the truth, since the assumption that I was lying leads to a contradiction. So, logic requires me to believe I was telling the truth. But at the time, I sure *felt* as though I was lying!

Speaking of lying, I must tell you the incident of Bertrand

Russell and the philosopher G. E. Moore. Russell described Moore as one of the most truthful people he had ever met. He once asked Moore, "Have you ever lied?" Moore replied, "Yes." In describing this incident, Russell wrote: "I think this is the only lie Moore ever told!"

The incident of my experience with the sales company raises the question of whether it is possible for a person to lie without knowing it. I would answer "No." To me, lying means making a statement, not which *is* false, but which one *believes* to be false. Indeed if a person makes a statement which happens to be true, but which he believes to be false, then I would say he is telling a lie.

I read of the following incident in a textbook on abnormal psychology. The doctors in a mental institution were thinking of releasing a certain schizophrenic patient. They decided to give him a lie-detector test. One of the questions they asked him was, "Are you Napoleon?" He replied, "No." The machine showed he was lying.

I also read somewhere the following incident showing how animals can sometimes dissimulate. An experiment was conducted with a chimpanzee in a room in which a banana was suspended by a string from the center of the ceiling. The banana was too high to reach. The room was empty except for the chimp, the experimenter, the banana and string, and several wooden boxes of various sizes. The purpose of the experiment was to determine whether the chimp was clever enough to make a scaffolding of the boxes, climb up, and reach the banana. What really happened was this: The experimenter stood in the corner of the room to watch the proceedings. The chimp came over to the corner and anxiously tugged the experimenter by the sleeve indicating that he wanted him to move. Slowly the experimenter followed the chimp. When they came to about the center of the room, the chimp suddenly jumped on his shoulders and got the banana.

3. The Joke Was on Me

A fellow graduate student of mine at the University of Chicago had two brothers, aged six and eight. I was a frequent visitor to their house and often did tricks for the children. One day I came and said, "I have a trick in which I could turn you both into lions." To my surprise, one of them said, "Okay, turn us into lions." I replied, "Well, uh, really, uh, I shouldn't do that, because there is no way I could turn you back again." The little one said, "I don't care; I want you to turn us into lions anyway." I replied, "No, really, there's *no* way I can turn you back." The older one shouted, "*I want you to turn us into lions!*" The little one then asked, "How do you turn us into lions?" I replied, "By saying the magic words." One of them asked, "What are the magic words?" I replied, "If I told you the magic words, I would be saying them, and so you would turn into lions." They thought about this for a while, and then one of them asked, "Aren't there any magic words which would bring us back?" I replied: "Yes, there are, but the trouble is this. If I said the first magic words, then not only you two but everybody in the world—including myself—would turn into a lion. And lions can't talk, so there would be no one left to say the other magic words to bring us back." The older one then said, "Write them down!" The little one said, "But I can't read!" I replied, "No, no, writing them down is out of the question; even if they were written down rather than said, everyone in the world would still turn into a lion." They said, "Oh."

About a week later I met the eight-year-old, and he said, "Smullyan, there's something I've been wanting to ask you; something which has been puzzling me." I replied, "Yes?" He said. "*How did you ever learn the magic words?*"

2. Puzzles and Monkey Tricks

SOME GOOD OLD-TIMERS

We will start with some good old-time puzzles which have amused many a generation. Some of these, many of you already know, but even for those in the know, I have a few new wrinkles.

4. Whose Picture Am I Looking At?

This puzzle was extremely popular during my childhood, but today it seems less widely known. The remarkable thing about this problem is that most people get the wrong answer but insist (despite all argument) that they are right. I recall one occasion about 50 years ago when we had some company and had an argument about this problem which seemed to last hours, and in which those who had the right answer just could not convince the others that they were right. The problem is this.

A man was looking at a portrait. Someone asked him, "Whose picture are you looking at?" He replied: "Brothers and sisters have I none, but this man's father is my father's son." ("This man's father" means, of course, the father of the man in the picture.)

Whose picture was the man looking at?

5. _____

Suppose, in the above situation, the man had instead answered: "Brothers and sisters have I none, but this man's son is my father's son." Now whose picture is the man looking at?

6. What Happens If an Irresistible Cannonball Hits an Immovable Post? _____

This is another problem from my childhood which I like very much. By an irresistible cannonball we shall mean a cannonball which knocks over everything in its way. By an immovable post we shall mean a post which cannot be knocked over by anything. So what happens if an irresistible cannonball hits an immovable post?

7. _____

The following is a very simple problem which many of you know. Twenty-four red socks and 24 blue socks are lying in a drawer in a dark room. What is the minimum number of socks I must take out of the drawer which will guarantee that I have at least two socks of the same color?

8. _____

A new twist on the above problem: Suppose some blue socks and the same number of red socks are in a drawer. Suppose it turns out that the minimum number of socks I must pick in order to be sure of getting at least one pair of the same color is the same as the minimum number I must pick in order to be sure of getting at least two socks of different colors. How many socks are in the drawer?

9. _____

Here is a well-known logic puzzle: Given that there are more inhabitants of New York City than there are hairs on the head of any inhabitant, and that no inhabitant is totally bald, does it necessarily follow that there must be at least two inhabitants with exactly the same number of hairs?

Here is a little variant of this problem: In the town of Podunk, the following facts are true:

(1) No two inhabitants have exactly the same number of hairs.
(2) No inhabitant has exactly 518 hairs.
(3) There are more imhabitants than there are hairs on the head of any one inhabitant.

What is the largest possible number of inhabitants of Podunk?

10. Who Was the Murderer? _____

This story concerns a caravan going through the Sahara desert. One night they pitched tents. Our three principle characters are A, B, and C. A hated C and decided to murder him by putting poison in the water of his canteen (this would be C's only water supply). Quite independently of this, B also decided to murder C, so (without realizing that C's water was already poisoned) he drilled a tiny hole in C's canteen so that the water would slowly leak out. As a result, several days later C died of thirst. The question is, who was the murderer, A or B? According to one argument, B was the murderer, since C never did drink the poison put in by A, hence he would have died even if A hadn't poisoned the water. According to the opposite argument, A was the real murderer, since B's actions had absolutely no effect on the outcome; once A poisoned the water, C was doomed, hence A would have died even if B had not drilled the hole.

Which argument is correct?

At this point I'll tell you the joke of a woodchopper from the Middle East who came looking for a job at a lumber camp.

The foreman said, "I don't know if this is the kind of job you want; here we chop trees." The woodchopper said, "That's precisely the sort of work I do." The foreman replied, "Okay, here's an axe—let's see how long it takes you to chop down this tree here." The woodchopper went over to the tree and felled it with one blow. The foreman, amazed, said, "Okay, try that big one over there." The woodchopper went over to the tree—biff, bam—in two strokes the tree was down. "Fantastic!" cried the foreman. "Of course you are hired, but how did you ever learn to chop like that?" "Oh," he replied, "I've had plenty of practice in the Sahara Forest." The foreman thought for a moment. "You mean," he said, "the Sahara Desert." "Oh yes," replied the woodchopper, "it is now!"

11. Another Legal Puzzle. _____

Two men were being tried for a murder. The jury found one of them guilty and the other one not guilty. The judge turned to the guilty one and said: "This is the strangest case I have ever come across! Though your guilt has been established beyond any reasonable doubts, the law compels me to set you free."
　　How do you explain this?

12. Two Indians. _____

Two American Indians were sitting on a log—a big Indian and a little Indian. The little Indian was the son of the big Indian, but the big Indian was not the father of the little Indian.
　　How do you explain that?

13. The Clock That Stopped. _____

Here is a cute simple old-time puzzle: A man owned no watch, but he had an accurate clock which, however, he sometimes forgot to wind. Once when this happened he went to the house of a friend, passed the evening with him, went back home, and set his clock. How could he do this without knowing beforehand the length of the trip?

14. Problem of the Bear. _____

The interesting thing about this problem is that many people have heard it and know the answer, but their reasons for the answer are insufficient. So even if you think you know the answer, be sure and consult the solution.

A man is 100 yards due south of a bear. He walks 100 yards due east, then faces due north, fires his gun due north, and hits the bear.

What color was the bear?

B. MONKEY TRICKS

At first I was undecided what title to give this book; I thought of "Recreational Logic," "Logical Recreations and Diversions," and others, but I was not too satisfied. Then I decided to consult *Roget's Thesaurus*: I looked in the index under "Recreations" and was directed to section 840 entitled "Amusement." There I came across such choice items as "fun," "frolic," "merriment," jollity," "heyday," "jocosity," "drollery," "buffoonery," "tomfoolery," "mummery." In the next paragraph I came across "play," "play at," "romps," "gambols," "pranks," "antic," "lark," "gambade," "*monkey trick*."[1] Well, when I saw "monkey trick," I laughed and said to my wife, "Hey, maybe I should call this book "Monkey Tricks." Delightful as that title is, however, it would have been misleading for this book as a whole, since most portions can hardly be described as monkey tricks. But the title serves perfectly for the items of this section, as the reader will soon realize.

[1] Italics mine.

15. **Problem of the Two Coins.** _____
Two American coins add up to thirty cents, yet one of them is not a nickel. What coins are they?

16. _____
Those of you who know anything about Catholicism, do you happen to know if the Catholic Church allows a man to marry his widow's sister?

17. _____
A man lived on the twenty-fifth floor of a thirty-story apartment building. Every morning (except Saturdays and Sundays) he got into the elevator, got off at the ground floor, and went to work. In the evening, he came home, got into the elevator, got off at the twenty-fourth floor, and walked up one flight.

Why did he get off at the twenty-fourth floor instead of the twenty-fifth?

18. **A Question of Grammar.** _____
Those of you who are interested in questions of good grammatical usage, is it more correct to say the yolk *is* white or the yolk *are* white?

19. **A Rate-Time Problem.** _____
A train leaves from Boston to New York. An hour later, a train leaves from New York to Boston. The two trains are going at exactly the same speed. Which train will be nearer to Boston when they meet?

20. **A Question of Slope.** _____
On a certain house, the two halves of the roof are unequally

pitched; one half slopes downward at an angle of 60° and the other half at an angle of 70°. Suppose a rooster lays an egg right on the peak. On which side of the roof would the egg fall?

21. How Many 9's? _____

A certain street contains 100 buildings. A sign-maker is called to number the houses from 1 to 100. He has to order numerals to do the job. Without using pencil and paper, can you figure out in your head how many 9's he will need?

22. The Racetrack. _____

A certain snail takes an hour and a half to crawl clockwise around a certain racetrack, yet when he crawls counter-clockwise around that same racetrack it takes him only ninety minutes. Why this discrepancy?

23. A Question of International Law. _____

If an airplane crashes right on the border of the United States and Canada, in which country would you bury the survivors?

24. How Do You Explain This? _____

A certain Mr. Smith and his son Arthur were driving in a car. The car crashed; the father was killed outright and the son Arthur was critically injured and rushed to a hospital. The old surgeon took a look at him and said, "I can't operate on him; he is my son Arthur!"

How do you explain this?

25. And Now! _____

And now, what is the name of this book?

4.

A remarkably large number of people arrive at the wrong answer that the man is looking at his own picture. They put themselves in the place of the man looking at the picture, and reason as follows: "Since I have no brothers or sisters, then my father's son must be me. Therefore I am looking at a picture of myself."

The first statement of this reasoning is absolutely correct; if I have neither brothers nor sisters, then my father's son is indeed myself. But it doesn't follow that "myself" is the answer to the problem. If the second clause of the problem had been, "this man is my father's son," then the answer to the problem would have been "myself." But the problem didn't say that; it said "this man's *father* is my father's son." From which it follows that this man's father is myself (since my father's son is myself). Since this man's father is myself, then I am this man's father, hence this man must be my son. Thus the correct answer to the problem is that the man is looking at a picture of his son.

If the skeptical reader is still not convinced (and I'm sure many of you are not!) it might help if you look at the matter a bit more graphically as follows:

(1) This man's father is my father's son.

Substituting the word "myself" for the more cumbersome phrase "my father's son" we get

(2) This man's father is myself.

Now are you convinced?

5.

The answer to the second problem, "Brothers and sisters

have I none, but this man's son is my father's son," is that the man is looking at a picture of his father.

6.

The given conditions of the problem are logically contradictory. It is logically impossible that there can exist *both* an irresistible cannonball *and* an immovable post. If an irresistible cannonball should exist, then by definition it would knock over any post in its way, hence there couldn't exist an immovable post. Alternatively, if there existed an immovable post, then by definition, no cannonball could knock it over, hence there could not exist an irresistible cannonball. Thus the existence of an irresistible cannonball is in itself not logically contradictory, nor is the existence of an immovable post in itself contradictory; but to assert they *both* exist is to assert a contradiction.

The situation is not really very different than had I asked you: "There are two people, John and Jack. John is taller than Jack and Jack is taller than John. Now, how do you explain *that?*" Your best answer would be, "Either you are lying, or you are mistaken."

7.

The most common wrong answer is "25." If the problem had been, "What is the smallest number I must pick in order to be sure of getting at least two socks of *different* colors," then the correct answer would have been 25. But the problem calls for at least two socks of the *same* color, so the correct answer is "three." If I pick three socks, then either they are all of the same color (in which case I certainly have at least two of the same color) or else two are of one color and the third is of the other color, so I then have two of the same color.

8.

The answer is four.

9.

In the first problem, the answer is "yes." For definiteness, assume there are exactly 8 million people in New York. If each inhabitant had a different number of hairs, then there would be 8 million different positive whole numbers each less than 8 million. This is impossible!

For the second problem, the answer is 518! To see this, suppose there were more than 518 inhabitants—say 520. Then there would have to be 520 distinct numbers all less than 520 and none of them equal to 518. This is impossible; there are exactly 520 distinct numbers (including zero) less than 520, hence there are only 519 numbers other than 518 which are less than 520.

Incidentally, one of the inhabitants of Podunk must be bald. Why?

10.

I doubt that either argument can precisely be called "correct" or "incorrect." I'm afraid that in a problem of this type, one man's opinion is as good as another's. My personal belief is that if anybody should be regarded as the cause of C's death, it was A. Indeed, if I were the defense attorney of B, I would point out to the court two things: (1) removing poisoned water from a man is in no sense killing him; (2) if anything, B's actions probably served only to prolong A's life (even though this was not his intention), since death by poisoning is likely to be quicker than death by thirst.

But then A's attorney could counter, "How can anyone in his right mind convict A of murder by poisoning when in fact C never drank any of the poison?" So, this problem is a real puzzler! It is complicated by the fact that it can be looked at from a moral angle, a legal angle, and a purely scientific angle involving the notion of *causation*. From a moral angle, obviously both men were guilty of *intent* to murder, but the sentence for actual murder is far more drastic. Regarding the legal angle, I do not know how the law would decide—perhaps different juries would decide differently. As for the scientific aspects of the problem, the whole notion of *causation* presents many problems. I think a whole book could be written on this puzzle.

11.

The two defendants were Siamese twins.

12.

The big Indian was the mother of the little Indian.

13.

When the man left his house he started the clock and jotted down the time it then showed. When he got to his friend's house he noted the time when he arrived and the time when he left. Thus he knew how long he was at his friend's house. When he got back home, he looked at the clock, so he knew how long he had been away from home. Subtracting from this the time he had spent at his friend's house, he knew how long the walk back and forth had been. Adding half of this to the time he left his friend's house, he then knew what time it really was now.

14.

The bear must be white; it must be a polar bear. The usual reason given is that the bear must have been standing at the North Pole. Well, this indeed is one possibility, but not the only one. From the North Pole, all directions are south, so if the bear is standing at the North Pole and the man is 100 yards south of him and walks 100 yards east, then when he faces north, he will be facing the North Pole again. But as I said, this is not the only solution. Indeed there is an infinite number of solutions. It could be, for example, that the man is very close to the South Pole on a spot where the Polar circle passing through that spot has a circumference of exactly 100 yards, and the bear is standing 100 yards north of him. Then if the man walks east 100 yards, he would walk right around that circle and be right back at the point he started from. So that is a second solution. But again, the man could be still a little closer to the South Pole at a point where the polar circle has a circumference of exactly 50 yards, so if he walked east 100 yards, he would walk around that little circle twice and be back where he started. Or he could be still a little closer to the South Pole at a point where the circumference of the polar circle is one-third of 100 yards, and walk east around the circle three times and be back where he started. And so forth for any positive integer n. Thus there is really an infinite number of places on the earth where the given conditions could be met.

Of course, in any of these solutions, the bear is sufficiently close to either the North Pole or the South Pole to qualify as a polar bear. There is, of course, the remote possibility that some mischievous human being deliberately transported a brown bear to the North Pole just to spite the author of this problem.

15

The answer is a quarter and a nickel. One of them (namely the quarter) is not a nickel.

16.

How can a dead man marry anybody?

17.

He was a midget and couldn't reach the elevator button for the twenty-fifth floor.

Someone I know (who is obviously not very good at telling jokes) once told this joke at a party at which I was present. He began thus: "On the twenty-fifth floor of an apartment building lived a midget...."

18.

Actually, the yolk is yellow.

19.

Obviously the two trains will be at the *same* distance from Boston when they meet.

20.

Roosters don't lay eggs.

21.

Twenty.

22.

There is no discrepancy; an hour and a half is the same as ninety minutes.

23.

One would hardly wish to bury the *survivors!*

24.

The surgeon was Arthur Smith's mother.

25.

Unfortunately, I cannot right now remember the name of this book, but don't worry, I'm sure it will come to me sooner or later.

3. Knights and Knaves

A. THE ISLAND OF KNIGHTS AND KNAVES

There is a wide variety of puzzles about an island in which certain inhabitants called "knights" always tell the truth, and others called "knaves" always lie. It is assumed that every inhabitant of the island is either a knight or a knave. I shall start with a well-known puzzle of this type and then follow it with a variety of puzzles of my own.

26.

According to this old problem, three of the inhabitants—A, B, and C—were standing together in a garden. A stranger passed by and asked A, "Are you a knight or a knave?" A answered, but rather indistinctly, so the stranger could not make out what he said. The stranger than asked B, "What did A say?" B replied, "A said that he is a knave." At this point the third man, C, said, "Don't believe B; he is lying!"

The question is, what are B and C?

27.

When I came upon the above problem, it immediately

struck me that C did not really function in any essential way; he was sort of an appendage. That is to say, the moment B spoke, one could tell without C's testimony that B was lying (see solution). The following variant of the problem eliminates that feature.

Suppose the stranger, instead of asking A what he is, asked A, "How many knights are among you?" Again A answers indistinctly. So the stranger asks B, "What did A say? B replies, "A said that there is one knight among us." Then C says, "Don't believe B; he is lying!"

Now what are B and C?

28. _____

In this problem, there are only two people, A and B, each of whom is either a knight or a knave. A makes the following statement: "At least one of us is a knave."

What are A and B?

29. _____

Suppose A says, "Either I am a knave or B is a knight." What are A and B?

30. _____

Suppose A says, "Either I am a knave or else two plus two equals five." What would you conclude?

31. _____

Again we have three people, A, B, C, each of whom is either a knight or a knave. A and B make the following statements:

    A: All of us are knaves.
    B: Exactly one of us is a knight.

What are A, B, C?

32.

Suppose instead, A and B say the following:

A: All of us are knaves.
B: Exactly one of us is a knave.

Can it be determined what B is? Can it be determined what C is?

33.

Suppose A says, "I am a knave, but B isn't."
What are A and B?

34.

We again have three inhabitants, A, B, and C, each of whom is a knight or a knave. Two people are said to be of the *same type* if they are both knights or both knaves. A and B make the following statements:

A: B is a knave.
B: A and C are of the same type.

What is C?

35.

Again three people A, B, and C. A says "B and C are of the same type." Someone then asks C, "Are A and B of the same type?"
What does C answer?

36. An Adventure of Mine.

This is an unusual puzzle; moreover it is taken from real life. Once when I visited the island of knights and knaves, I came across two of the inhabitants resting under a tree. I asked one of them, "Is either of you a knight?" He responded, and I knew the answer to my question.

What is the person to whom I addressed the question—is he a knight or a knave; And what is the other one? I can assure you, I have given you enough information to solve this problem.

37.

Suppose *you* visit the island of knights and knaves. You come across two of the inhabitants lazily lying in the sun. You ask one of them whether the other one is a knight, and you get a (yes-or-no) answer. Then you ask the second one whether the first one is a knight. You get a (yes-or-no) answer.

Are the two answers necessarily the same?

38. Edward or Edwin?

This time you come across just one inhabitant lazily lying in the sun. You remember that his first name is either Edwin or Edward, but you cannot remember which. So you ask him his first name and he answers "Edward."

What is his first name?

B. KNIGHTS, KNAVES, AND NORMALS

An equally fascinating type of problem deals with three types of people: knights, who always tell the truth; knaves, who always lie; and normal people, who sometimes lie and sometimes tell the truth. Here are some puzzles of mine about knights, knaves, and normals.

39.

We are given three people, A,B,C, one of whom is a knight,

one a knave, and one normal (but not necessarily in that order). They make the following statements:

> A: I am normal.
> B: That is true.
> C: I am not normal.

What are A, B, and C?

40. _____

Here is an unusual one: Two people, A and B, each of whom is either a knight, or knave, or a normal, make the following statements:

> A: B is a knight.
> B: A is not a knight.

Prove that at least one of them is telling the truth, but is not a knight.

41 _____

This time A and B say the following:

> A: B is a knight.
> B: A is a knave.

Prove that either one of them is telling the truth but is not a knight, or one of them is lying but is not a knave.

42. A Matter of Rank. _____

On this island of knights, knaves, and normals, knaves are said to be of the *lowest* rank, normals of *middle* rank, and knights of *highest* rank.

I am particularly partial to the following problem: Given two people A,B, each of whom is a knight, a knave, or a normal, they make the following statements:

A: I am of lower rank than B.
B: That's not true!

Can the ranks of either A or B be determined? Can it be determined of either of these statements whether it is true or false?

43. _____

Given three people A,B,C, one of whom is a knight, one a knave, and one normal. A,B, make the following statements:

A: B is of higher rank than C.
B: C is of higher rank than A.

Then C is asked: "Who has higher rank, A or B?" What does C answer?

C. THE ISLAND OF BAHAVA

The island of Bahava is a female liberationist island; hence the women are also called *knights, knaves,* or *normals.* An ancient empress of Bahava once, in a whimsical moment, passed a curious decree that a knight could marry only a knave and a knave could marry only a knight. (Hence a normal can marry only a normal.) Thus, given any married couple, either they are both normal, or one of them is a knight and the other a knave.

The next three stories all take place on the island of Bahava.

44. _____

We first consider a married couple, Mr. and Mrs. A. They make the following statements:

Mr. A / My wife is not normal.
Mrs. A / My husband is not normal.

What are Mr. and Mrs. A?

45.

Suppose, instead, they had said:

> *Mr. A* / My wife is normal.
> *Mrs. A* / My husband is normal.

Would the answer have been different?

46.

This problem concerns two married couples on the island of Bahava, Mr. and Mrs. A, and Mr. and Mrs. B. They are being interviewed, and three of the four people give the following testimony:

> *Mr. A* / Mr. B is a knight.
> *Mrs. A* / My husband is right; Mr. B is a knight.
> *Mrs. B* / That's right. My husband is indeed a knight.

What are each of the four people, and which of the three statements are true?

SOLUTIONS

26.

It is impossible for either a knight or a knave to say, "I'm a knave," because a knight wouldn't make the false statement that he is a knave, and a knave wouldn't make the true statement that he is a knave. Therefore A never did say that he was a knave. So B lied when he said that A said that he was a knave. Hence B is a knave. Since C said that B was lying and B was indeed lying, then C spoke the truth, hence is a knight. Thus B is a knave and C is a knight. (It is impossible to know what A is.)

27.

The answer is the same as that of the preceding problem, though the reasoning is a bit different.

The first thing to observe is that B and C must be of opposite types, since B contradicts C. So of these two, one is a knight and the other a knave. Now, if A were a knight, then there would be two knights present, hence A would not have lied and said there was only one. On the other hand, if A were a knave, then it would be true that there was exactly one knight present; but then A, being a knave, couldn't have made that true statement. Therefore A could not have said that there was one knight among them. So B falsely reported A's statement, and thus B is a knave and C is a knight.

28.

Suppose A were a knave. Then the statement "At least one of us is a knave" would be false (since knaves make false statements); hence they would both be knights. Thus, if A were a knave he would also have to be a knight, which is impossible. Therefore A is not a knave; he is a knight. Therefore his statement must be true, so at least one of them really is a knave. Since A is a knight, then B must be the knave. So A is a knight and B is a knave.

29.

This problem is a good introduction to the logic of disjunction. Given any two statements p, q, the statement "either p or q" means that at least one (and possibly both) of the statements p, q are true. If the statement "either p or q" should be false, then *both* the statements p, q are false. For example, if I should say, "Either it is raining or it is snowing," then if my statement is incorrect, it is both false that it is raining and false that it is snowing.

This is the way "either/or" is used in logic, and is the way it will be used throughout this book. In daily life, it is sometimes used this way (allowing the possibility that both alternatives hold) and sometimes in the so-called "exclusive" sense—that one and only one of the conditions holds. As an example of the exclusive use, if I say; "I will marry Betty or I will marry Jane," it is understood that the two possibilities are mutually exclusive—that is, that I will not marry both girls. On the other hand, if a college catalogue states that an entering student is required to have had *either* a year of mathematics *or* a year of a foreign language, the college is certainly not going to exclude you if you had both! This is the "*inclusive*" use of "either/or" and is the one we will constantly employ.

Another important property of the disjunction relation "either this or that" is this. Consider the statement "*p* or *q*" (which is short for "either *p* or *q*"). Suppose the statement happens to be true. Then if *p* is false, *q* must be true (because at least one of them is true, so if *p* is false, *q* must be the true one). For example, suppose it is true that it is either raining or snowing, but it is false that it is raining. Then it must be true that it is snowing.

We apply these two principles as follows. A made a statement of the disjunctive type: "Either I am a knave or B is a knight." Suppose A is a knave. Then the above statement must be false. This means that it is neither true that A is a knave nor that B is a knight. So if A were a knave, then it would follow that he is not a knave—which would be a contradiction. Therefore A must be a knight.

We have thus established that A is a knight. Therefore his statement is true that at least one of the possibilities holds: (1) A is a knave; (2) B is a knight. Since possibility (1) is false (since A is a knight) then possibility (2) must be the correct one, i.e., B is a knight. Hence A,B, are both knights.

30.

The only valid conclusion is that the author of this problem is not a knight. The fact is that neither a knight nor a knave could possibly make such a statement. If A were a knight, then the statement that either A is a knave or that two plus two equals five would be false, since it is neither the case that A is a knave nor that two plus two equals five. Thus A, a knight, would have made a false statement, which is impossible. On the other hand, if A were a knave, then the statement that either A is a knave or that two plus two equals five would be true, since the first clause that A is a knave is true. Thus A, a knave, would have made a true statement, which is equally impossible.

Therefore the conditions of the problem are contradictory (just like the problem of the irresistible cannonball and the immovable post). Therefore, I, the author of the problem, was either mistaken or lying. I can assure you I wasn't mistaken. Hence it follows that I am not a knight.

For the sake of the records, I would like to testify that I have told the truth at least once in my life, hence I am not a knave either.

31.

To begin with, A must be a knave, for if he were a knight, then it would be true that all three are knaves and hence that A too is a knave. If A were a knight he would have to be a knave, which is impossible. So A is a knave. Hence his statement was false, so in fact there is at least one knight among them.

Now, suppose B were a knave. Then A and B would both be knaves, so C would be a knight (since there is at least one knight among them). This would mean that there was exactly one knight among them, hence B's statement would be true. We would thus have the impossibility of a knave making a true statement. Therefore B must be a knight.

We now know that A is a knave and that B is a knight. Since B is a knight, his statement is true, so there is exactly one knight among them. This knight must be B, hence C must be a knave. Thus the answer is that A is a knave, B is a knight, and C is a knave.

32.

It cannot be determined what B is, but it can be proved that C is a knight.

To begin with, A must be a knave for the same reasons as in the preceding problem; hence also there is at least one knight among them. Now, either B is a knight or a knave. Suppose he is a knight. Then it is true that exactly one of them is a knave. This only knave must be A, so C would be a knight. So if B is a knight, so is C. On the other hand, if B is a knave, then C must be a knight, since all three can't be knaves (as we have seen). So in either case, C must be a knight.

33.

To begin with, A can't be a knight or his statement would be true, in which case he would have to be a knave. Therefore A is a knave. Hence also his statement is false. If B were a knight, then A's statement would be true. Hence B is also a knave. So A,B are both knaves.

34.

Suppose A is a knight. Then his statement that B is a knave must be true, so B is then a knave. Hence B's statement that A and C are of the same type is false, so A and C are of different types. Hence C must be a knave (since A is a knight). Thus if A is a knight, then C is a knave.

On the other hand, suppose A is a knave. Then his statement that B is a knave is false, hence B is a knight.

Hence B's statement is true that A and C are of the same type. This means that C must be a knave (since A is).

We have shown that regardless of whether A is a knight or a knave, C must be a knave. Hence C is a knave.

35.

I'm afraid we can solve this problem only by analysis into cases.

Case One: A is a knight. Then B,C really are of the same type. If C is a knight, then B is also a knight, hence is of the same type as A, so C being truthful must answer "Yes." If C is a knave, then B is also a knave (since he is the same type as C), hence is of a different type than A. So C, being a knave, must lie and say "Yes."

Case Two: A is a knave. Then B,C are of different types. If C is a knight, then B is a knave, hence he is of the same type as A. So C, being a knight, must answer "Yes." If C is a knave, then B, being of a different type than C, is a knight, hence is of a different type than A. Then C, being a knave, must lie about A and C being of different types, so he will answer "Yes."

Thus in both cases, C answers "Yes."

36.

To solve this problem, you must use the information I gave you that after the speaker's response, I knew the true answer to my question.

Suppose the speaker—call him A—had answered "Yes." Could I have then known whether at least one of them was a knight? Certainly not. For it could be that A was a knight and truthfully answered "Yes" (which would be truthful, since at least one—namely A—was a knight), or it could be that both of them were knaves, in which case A would have falsely answered "Yes" (which would indeed be false since neither was a knight). So if A had answered "Yes" I would have had no way of knowing. But I told you that I *did* know after A's answer. Therefore A must have answered "No."

The reader can now easily see what A and the other—call him B—must be: If A were a knight, he couldn't have truthfully answered "No," so A is a knave. Since his answer "No" is false, then there is at least one knight present. Hence A is a knave and B is a knight.

37.

Yes, they are. If they are both knights, then they will both answer "Yes." If they are both knaves, then again they will both answer "Yes." If one is a knight and the other a knave, then the knight will answer "No," and the knave will also answer "No."

38.

I feel entitled, occasionally, to a little horseplay. The vital clue I gave you was that the man was lazily lying in the sun. From this it follows that he was lying in the sun. From this it follows that he was *lying*, hence he is a knave. So his name is Edwin.

39.

To begin with, A cannot be a knight, because a knight would never say that he is normal. So A is a knave or is normal. Suppose A were normal. Then B's statement would be true, hence B is a knight or a normal, but B can't be normal (since A is), so B is a knight. This leaves C a knave. But a knave cannot say that he is not normal (because a knave really isn't normal), so we have a contradiction. Therefore A cannot be normal. Hence A is a knave. Then B's statement is false, so B must be normal (he can't be a knave since A is). Thus A is the knave, B is the normal one, hence C is the knight.

40.

The interesting thing about this problem is that it is impossible to know whether it is A who is telling the truth but isn't a knight or whether it is B who is telling the truth but isn't a knight; all we can prove is that at least one of them has that property.

Either A is telling the truth or he isn't. We shall prove: (1) If he is, then A is telling the truth but isn't a knight; (2) If he isn't, then B is telling the truth but isn't a knight.

(1) Suppose A is telling the truth. Then B really is a knight. Hence B is telling the truth, so A isn't a knight. Thus if A is telling the truth then A is a person who is telling the truth but isn't a knight.

(2) Suppose A is not telling the truth. Then B isn't a knight. But B must be telling the truth, since A can't be a knight (because A is not telling the truth). So in this case B is telling the truth but isn't a knight.

41.

We shall show that if B is telling the truth then he isn't a knight, and if he isn't telling the truth then A is lying but isn't a knave.

(1) Suppose B is telling the truth. Then A is a knave, hence A is certainly not telling the truth, hence B is not a knight. So in this case B is telling the truth but isn't a knight.

(2) Suppose B is not telling the truth. Then A is not really a knave. But A is certainly lying about B, because B can't be a knight if he isn't telling the truth. So in this case, A is lying but isn't a knave.

42.

To begin with, A can't be a knight, because it can't be true that a knight is of lower rank than anyone else. Now, suppose A is a knave. Then his statement is false, hence he is not of lower rank than B. Then B must also be a knave (for if he weren't, then A *would* be of lower rank than B). So if A is a knave, so is B. But this is impossible because B is contradicting A, and two contradictory claims can't both be false. Therefore the assumption that A is a knave leads to a contradiction. Therefore A is not a knave. Hence A must be normal.

Now, what about B? Well, if he were a knight, then A (being normal) actually would be of lower rank than B, hence A's statement would be true, hence B's statement false, and we would have the impossibility of a knight making a false statement. Thus B is not a knight. Suppose B were a knave. Then A's statement would be false, hence B's would be true, and we would have a knave making a true statement. Therefore B can't be a knave either. Hence B is normal.

Thus A and B are both normal. So also, A's statement is false and B's statement is true. So the problem admits of a complete solution.

43. _____

Step 1: We first show that from A's statement if follows that C cannot be normal. Well, if A is a knight then B really is of higher rank than C, hence B must be normal and C must be a knave. So in this case, C is not normal. Suppose A is a knave. Then B is not really of higher rank than C, hence B is of lower rank, so B must be normal and C must be a knight. So in this case, C again is not normal. The third possible case is that A is normal, in which case C certainly isn't (since only one of A, B, C is normal). Thus C is not normal.

Step 2: By similar reasoning, it follows from B's statement that A is not normal. Thus neither A nor C is normal. Therefore B is normal.

Step 3: Since C is not normal, then he is a knight or a knave. Suppose he is a knight. Then A is a knave (since B is normal) hence B is of higher rank than A. So C, being a knight, would truthfully answer, "B is of higher rank." On the other hand, suppose C is a knave. Then A must be a

---

knight, so B is not of higher rank than A. Then C, being a knave, would lie and say, "B is of higher rank than A." So regardless of whether C is a knight or a knave, he answers that B is of higher rank than A.

44.

Mr. A cannot be a knave, because then his wife would be a knight and hence not normal, so Mr. A's statement would have been true. Similarly Mrs. A cannot be a knave. Therefore neither is a knight either (or the spouse would then be a knave), so they are both normal (and both lying).

45.

For the second problem, the answer is the same. Why?

46.

It turns out that all four are normal, and all three statements are lies.

First of all, Mrs. B must be normal, for if she were a knight her husband would be a knave, hence she wouldn't have lied and said he was a knight. If she were a knave, her husband would be a knight, but then she wouldn't have told the truth about this. Therefore Mrs. B is normal. Hence also Mr. B is normal. This means that Mr. and Mrs. A were both lying. Therefore neither one is a knight, and they can't both be knaves, so they are both normal.

4. Alice in the Forest of Forgetfulness

A. THE LION AND THE UNICORN

When Alice entered the Forest of Forgetfulness, she did not forget *everything*; only certain things. She often forgot her name, and the one thing she was most likely to forget was the day of the week. Now, the Lion and the Unicorn were frequent visitors to the forest. These two are strange creatures. The Lion lies on Mondays, Tuesdays, and Wednesdays and tells the truth on the other days of the week. The Unicorn, on the other hand, lies on Thursdays, Fridays, and Saturdays, but tells the truth on the other days of the week,

47.

One day Alice met the Lion and the Unicorn resting under a tree. They made the following statements:

Lion / Yesterday was one of my lying days.
Unicorn / Yesterday was one of my lying days too.

From these two statements, Alice (who was a very bright girl) was able to deduce the day of the week. What day was it?

48.

On another occasion Alice met the Lion alone. He made the following two statements:

(1) I lied yesterday.
(2) I will lie again two days after tomorrow.

What day of the week was it?

49.

On what days of the week is it possible for the Lion to make the following two statements:

(1) I lied yesterday.
(2) I will lie again tomorrow.

50.

On what days of the week is it possible for the Lion to make the following single statement: "I lied yesterday and I will lie again tomorrow." *Warning!* The answer is *not* the same as that of the preceding problem!

B. TWEEDLEDUM AND TWEEDLEDEE

During one month the Lion and the Unicorn were absent from the Forest of Forgetfulness. They were elsewhere, busily fighting for the crown.

However, Tweedledum and Tweedledee were frequent visitors to the forest. Now, one of the two is like the Lion, lying on Mondays, Tuesdays, and Wednesdays and telling the truth on the other days of the week. The other one is like the Unicorn; he lies on Thursdays, Fridays, and Saturdays but tells the truth the other days of the week. Alice didn't know which one was like the Lion and which one was like the Unicorn. To make matters worse, the brothers looked so much alike, that Alice could not even tell them apart (except when they wore their embroidered collars, which they seldom did). Thus poor Alice found the situation most confusing indeed! Now, here are some of Alice's adventures with Tweedledum and Tweedledee.

51. _____

One day Alice met the brothers together and they made the following statements:

> *First One* / I'm Tweedledum.
> *Second One* / I'm Tweedledee.

Which one was really Tweedledum and which one was Tweedledee?

52. _____

On another day of that same week, the two brothers made the following statements:

> *First One* / I'm Tweedledum.
> *Second One* / If that's really true, then I'm Tweedledee!

Which was which?

53. _____

On another occasion, Alice met the two brothers, and asked one of them, "Do you lie on Sundays?" He replied "Yes." Then she asked the other one the same question. What did he answer?

54.

On another occasion, the brothers made the following
statements:

> *First One* / (1) I lie on Saturdays.
> (2) I lie on Sundays.
> *Second One* / I will lie tomorrow.

What day of the week was it?

55.

One day Alice came across just one of the brothers. He
made the following statement: "I am lying today and I am
Tweedledee."
    Who was speaking?

56.

Suppose, instead, he had said: "I am lying today or I am
Tweedledee." Would it have been possible to determine
who it was?

57.

One day Alice came across both brothers. They made the
following statements:

> *First One* / If I'm Tweedledum then he's Tweedledee.
> *Second One* / If he's Tweedledee then I'm
> Tweedledum.

Is it possible to determine who is who? Is it possible to
determine the day of the week?

58. A Mystery Resolved! _____

On this great occasion, Alice resolved three grand mysteries. She came across the two brothers grinning under a tree. She hoped that on this encounter she would find out three things: (1) the day of the week; (2) which of the two was Tweedledum; (3) whether Tweedledum was like the Lion or the Unicorn in his lying habits (a fact she had long desired to know!)

Well, the two brothers made the following statements:

First One / Today is not Sunday.
Second One / In fact, today is Monday.
First One / Tomorrow is one of Tweedledee's lying days.
Second One / The Lion lied yesterday.

Alice clapped her hands in joy. The problem was now completely solved. What is the solution?

C. WHO OWNS THE RATTLE?

Tweedledum and Tweedledee
    Agreed to have a battle;
For Tweedledum said Tweedledee
    Had spoiled his nice new rattle.

Just then flew down a monstrous crow,
    As black as a tar-barrel,
Which frightened both the heroes so
    They quite forgot their quarrel.
                    —Old Nursery Rhyme

"Well, well," triumphantly exclaimed the White King to Alice one day, "I've found the rattle, and I've had it restored. Doesn't it look as good as new?"

"Yes, indeed," replied Alice, admiringly, "it looks as good as the day it was made. Even a baby couldn't tell the difference."

"What do you mean *even* a baby?" cried the White King sternly. "That not very logical, you know. Of course a baby couldn't tell the difference—one would hardly expect a baby to do that!"

"What you *should* have said," continued the King, somewhat more gently, "is that even a grownup couldn't tell the difference—not even the world's greatest rattle expert."

"Anyway," continued the King, "we'll imagine it said. The important thing is to restore the rattle to its rightful owner. Will you please do this for me?"

"Who is the rightful owner?" asked Alice.

"I shouldn't have to tell you *that!*" cried the King impatiently.

"Why not?" inquired Alice.

"Because it says quite explicitly in the rhyme—which I'm sure you know—that Tweedledum said that Tweedledee had spoiled his nice new rattle, so the rattle belongs to Tweedledum, of course!"

"Not necessarily," replied Alice, who was in a mood for a little argument, "I know the rhyme well, and I believe it."

"Then what's the problem?" cried the King, more puzzled than ever.

"Very simple, really," explained Alice. "I grant that what the rhyme says is true. Therefore Tweedledum did indeed *say* that Tweedledee had spoiled his rattle. But because Tweedledum said it, it does not mean that it is necessarily true. Perhaps Tweedledum said it on one of his lying days. Indeed, for all I know, it may be the other way around—maybe it was Tweedledum who spoiled Tweedledee's new rattle."

"Oh, dear," replied the King disconsolately, "I never thought of that. Now all my good intentions are wasted."

The poor king looked so dejected, Alice thought he would cry. "Never mind," said Alice as cheerfully as she could. "Give me the rattle and I will try to find out who is the true owner. I've had some experience with liars and truth-tellers around here, and I have gotten a little of the knack of how to handle them."

"I hope so!" replied the King mournfully.

Now I shall tell you of Alice's actual adventures with the rattle.

59. _____

She took the rattle and went into the Forest of Forgetfulness, hoping to find at least one of the brothers. To her great delight, she suddenly came across both of them grinning under a tree. She went to the first one and sternly said: "I want the truth now! Who really owns the rattle?" He replied, "Tweedledee owns the rattle." She thought for a while, and asked the second one, "Who are you?" He replied, "Tweedledee."

Now, Alice did not remember the day of the week, but she was sure it was not Sunday.

To whom should Alice give the rattle?

60. _____

Alice restored the rattle to its rightful owner. Several days later, the other brother broke the rattle again. This time, no black crow came to frighten the brothers, so they began slamming and banging away at each other. Alice picked up the broken rattle and ran out of the forest as fast as she could.

Some time later, she again came across the White King. She thoroughly explained the situation to him.

"Very interesting," replied the King. "The most remarkable part is that although you knew to whom to give it, we still do not know if it is Tweedledee or Tweedledum who owns the rattle."

"Very true," replied Alice, "but what do I do now?"

"No problem," replied the King, I can easily have the rattle fixed again."

True to his word, the White King had the rattle

perfectly restored and gave it to Alice some days later. Alice went trepidly into the forest, fearing that the battle might still be on. As a matter of fact, the brothers had called a temporary truce, and Alice came across just one of them resting wearily under a tree. Alice went over to him and asked, "Who really owns this rattle?" He quizzically replied, "The true owner of this rattle is lying today."

What are the chances that the speaker owns the rattle?

61.

Several days later Alice again came upon just one of the brothers lying under a tree. She asked the same question, and the reply was, "The owner of this rattle is telling the truth today."

Alice pondered over this; she wondered just what were the chances that the speaker owned the rattle.

"I know what you are thinking," said Humpty Dumpty, who happened to be standing nearby, "and the chances are exactly thirteen out of fourteen!"

How did Humpty Dumpty ever arrive at those numbers?

62.

This time Alice came across both brothers together. Alice asked the first one, "Do you own this rattle?" He replied "Yes." Then Alice asked the second one, "Do *you* own this rattle?" The second one answered, and Alice gave one of them the rattle.

Did Alice give the rattle to the first or the second one?

D. FROM THE MOUTH OF THE JABBERWOCKY

Of all the adventures Alice had with the Tweedle brothers in the Forest of Forgetfulness, the one I am about to relate was the most eerie, and the one Alice remembered most vividly.

It started this way: One day Humpty Dumpty met Alice and said: "Child, I wish to tell you a great secret. Most people don't know it, but Tweedledee and Tweedledum actually have a third brother—his name is Tweedledoo. He lives in a far-off land but occasionally comes around to these parts. He looks as much like Tweedledee and Tweedledum as Tweedledee and Tweedledum look like each other."

This information disturbed Alice dreadfully! For one thing, the possibility that there really was a third one would mean that all her past inferences were invalidated, and that she really may not have figured out the day of the week when she thought she had. Of even greater practical importance, she may not have restored the rattle to its rightful owner after all.

Alice pondered deeply over these troublesome thoughts Finally, she asked Humpty Dumpty a sensible question.

"On what days does Tweedledoo lie?"

"Tweedledoo always lies," replied Humpty Dumpty.

Alice walked away in troubled silence. "Perhaps the whole thing is only a fabrication of Humpty Dumpty," Alice thought to herself. "It certainly sounds a most unlikely tale to me." Still, Alice was haunted by the thought that it *might* be true.

There are four different accounts of just what happened next, and I shall tell you all of them. I ask the reader to assume two things: (1) if there really is an individual other than Tweedledee or Tweedledum who looks indistinguishable from them, then his name really is Tweedledoo; (2) if such an individual exists, then he really does lie all the time. I might remark that the second assumption is not necessary for the solution of the next mystery, but it is for the two which follow after that.

63. The First Version. _____

Alice came across just one brother alone in the forest. At least, he *looked* like he was Tweedledee or Tweedledum.

Alice told him Humpty Dumpty's story, and then asked him, "Who are you really?" He gave the enigmatic reply, "I am either Tweedledee or Tweedledum, and today is one of my lying days."

The question is, does Tweedledoo really exist, or is he just a fabrication of Humpty Dumpty?

64. The Second Version.

According to this version, Alice came across (what seemed to be) both brothers. She asked the first one: "Who really are you?" She got the following replies:

First One / "I'm Tweedledoo."
Second One / "Yes, he is!"

What do you make of this version?

65. The Third Version.

According to this version, Alice came across just one of them. He made the following statement: "Today is one of my lying days." What do you make of this version?

66. The Fourth Version.

According to this version, Alice met (what seemed to be) both brothers on a weekday. She asked, "Does Tweedledoo really exist?" She got the following replies:

First One / Tweedledoo exists.
Second One / I exist.

What do you make of this version?

Epilogue.

Now, what is the real truth of the matter; does Tweedledoo really exist or not? Well, I have given you four conflicting versions of what really happened. How come four versions? Well, to tell you the truth, I didn't invent these stories myself; I heard them all from the mouth of the Jabberwocky. Now, the conversation between Alice and Humpty Dumpty really happened: Alice told me this herself, and Alice is always truthful. But the four versions of what happened after that were all told to me by the Jabberwocky. Now, I know that the Jabberwocky lies on the same days as the Lion (Monday, Tuesday, Wednesday) and he told me these stories on four consecutive weekdays. (I know they were weekdays, because I am lazy and sleep all day Saturdays and Sundays.) They were told to me in the same order as I recounted them.

From this information, the reader should have no difficulty in ascertaining whether Tweedledoo really exists or whether Humpty Dumpty was lying. Does Alice know whether Tweedledoo exists?

SOLUTIONS

47.

The only days the Lion can say "I lied yesterday" are Mondays and Thursdays. The only days the Unicorn can say "I lied yesterday" are Thursdays and Sundays. Therefore the only day they can both say that is on Thursday.

48.

The lion's first statement implies that it is Monday or Thursday. The second statement implies that it is not Thursday. Hence it is Monday.

49.

On no day of the week is this possible! Only on Mondays and Thursdays could he make the first statement; only on

Wednesdays and Sundays could he make the second. So there is no day he could say both.

50. _____

This is a very different situation! It well illustrates the difference between making two statements separately and making one statement which is the conjunction of the two. Indeed, given any two statements X, Y, if the single statement "X and Y" is true, then it of course follows that X, Y are true separately; but if the conjunction "X and Y" is false, it only follows that at least one of them is false.

Now, the only day of the week it could be true that the Lion lied yesterday and will lie again tomorrow is Tuesday (this is the one and only day which occurs between two of the Lion's lying days). So the day the Lion said that couldn't be Tuesday, for on Tuesdays that statement is true, but the Lion doesn't make true statements on Tuesdays. Therefore it is not Tuesday, hence the Lion's statement is false, so the Lion is lying. Therefore the day must be either Monday or Wednesday.

51. _____

If the first statement is true, then the first one really is Tweedledum, hence the second one is Tweedledee and the second statement is also true. If the first statement is false, then the first one is actually Tweedledee and the second one is Tweedledum, and hence the second statement is also false. Therefore either both statements are true or both statements are false. They can't both be false, since the brothers never lie on the same day. Therefore both statements must be true. So the first one is Tweedledum and the second one is Tweedledee. Also, the day of the encounter must be Sunday.

52.

This is a horse of a very different color! The second one's statement is certainly true. Now, we are given that the day of the week is different from that of the last problem, so it is a weekday. Therefore it cannot be that both statements are true, so the first one must be false. Therefore the first one is Tweedledee and the second is Tweedledum.

53.

The first answer was clearly a lie, hence the event must have taken place on a weekday. Therefore the other one must have answered truthfully and said "No."

54.

Statement (2) of the first one is clearly false, hence statement (1) is false too (since it is uttered on the same day). Therefore the first one does not lie on Saturdays, so the second one lies on Saturdays. The second one is telling the truth on this day (since the first one is lying), so it is now Monday, Tuesday, or Wednesday. The only one of these days in which it is true that he will lie tomorrow is Wednesday. So the day is Wednesday.

55.

His statement is certainly false (for if it were true, then he would be lying today, which is a contradiction). Therefore at least one of the two clauses "I am lying today," "I am Tweedledee" must be false. The first clause ("I am lying today") is true, therefore the second clause must be false. So he is Tweedledum.

56.

Yes it would. If he were lying today, then the first clause of the disjunction would be true, hence the whole statement would be true, which is a contradiction. Therefore he is telling the truth today. So his statement is true: either he is lying today or he is Tweedledee. Since he is not lying today, then he is Tweedledee.

57.

Both statements are obviously true, so it is a Sunday. It is not possible to determine who is who.

58.

To begin with, it is impossible on a Sunday for either brother to lie and say that it is not Sunday. Therefore today cannot be Sunday. So the first one is telling the truth, and (since it is not Sunday), the second one is therefore lying today. The second one says today is Monday, but he is lying, so it is not Monday either.

Now, the second one has also told the lie that the Lion lied yesterday, hence yesterday was really one of the Lion's truthful days. This means that yesterday was Thursday, Friday, Saturday, or Sunday, so today is Friday, Saturday, Sunday, or Monday. We have already ruled out Sunday and Monday, so today must be Friday or Saturday.

Next we observe that tomorrow is one of Tweedledee's lying days (since the first one, who is speaking the truth, said so). Therefore today cannot be Saturday. Hence today is Friday.

From this it further follows that Tweedledee lies on Saturdays, hence he is like the Unicorn. Also, the first one is telling the truth today, which is a Friday, hence he is Tweedledum. This proves everything.

59.

Suppose the first one told the truth. Then the rattle belongs to Tweedledee. The second speaker must be lying (since it is not Sunday), hence his name is not really Tweedledee; it is Tweedledum. Hence the first speaker is Tweedledee and should get the rattle.

Suppose the first one lied. Then the rattle belongs to Tweedledum. Then also the second one told the truth so is really Tweedledee. Then again the first one owns the rattle. So in either case, the rattle belongs to the first speaker.

60.

The chances are zero! Suppose his statement is true. Then the owner of the rattle is lying today, hence cannot be the speaker. Suppose on the other hand that his statement is false. Then the owner of the rattle is telling the truth today, hence again cannot be the speaker.

61.

Humpty Dumpty was right! Suppose the speaker is lying. Then the owner of the rattle is not telling the truth today; he is lying today, hence must be the speaker. But suppose the speaker is telling the truth. Then the owner of the rattle is indeed telling the truth today. If it is a weekday, then he must be the owner, but if it is a Sunday, then both brothers are telling the truth today, so either could be the owner.

In summary, if it is a weekday, then the speaker is definately the owner. If it is Sunday, then the chances are even that he is the owner. Therefore the chances are 6½ out of 7—or 13 out of 14—that he is the owner.

62.

The clue here is that Alice *did* know who to give it to. Had the second one answered "Yes," then one of them would have been telling the truth and the other lying, hence Alice would have no way of knowing who owned the rattle. But I told you she did know, hence the second one didn't answer "Yes." Therefore they were both lying or both telling the truth. This means they were both telling the truth, and it must have been Sunday. So Alice gave it to the first one.

63.

Yes, Tweedledoo must exist; Alice was just talking to him.

The speaker claimed that the following statements are *both* true:

(1) He is either Tweedledee or Tweedledum
(2) He is lying today.

If his claim were true, then (1) and (2) would both be true, hence (2) would be true, which would be a contradiction. Therefore his claim is false, so (1) and (2) cannot both be true. Now, (2) is true (since his claim on this day is false), so it must be (1) that is not true. Therefore he is neither Tweedledee nor Tweedledum, so he must be Tweedledoo.

64.

The first one can't really be Tweedledoo (since Tweedledoo always lies); so he is Tweedledee or Tweedledum, but he is lying. Then the second one is also lying. If the second one were Tweedledee or Tweedledum, then Tweedledee and Tweedledum would be lying on the same day, which is impossible. Therefore the second one must be Tweedledoo.

65.

This version is just simply false!

66.

Whoever the second one is, his statement is certainly true. (I think Descartes pointed out that anyone who says he exists is making a true statement; certainly I have never met anyone who didn't exist.) Since the second statement is true and it is not Sunday, then the first statement must be false.

So if this version of the story is correct, Tweedledoo doesn't exist.

Solution to the Epilogue. _____

The third version of the story is definitely false. Also none of the stories was told on a Saturday or Sunday. The only way these four stories can be fitted into four consecutive days satisfying these conditions is that the third version was told on a Wednesday. So the last version was told on a Thursday, hence must be the true one. So Tweedledoo doesn't really exist! (I'm quite sure, incidentally, that had Tweedledoo really existed, Lewis Carroll would have known about it.)

As for Alice, since the fourth version is the only one which really took place, then Alice should have no difficulty in realizing that all these "Tweedledoo fears" were groundless.

PART TWO

Portia's Caskets and Other Mysteries

5. The Mystery of Portia's Caskets

A. THE FIRST TALE

67a.

In Shakespeare's *Merchant of Venice* Portia had three caskets—gold, silver, and lead—inside one of which was Portia's portrait. The suitor was to choose one of the caskets, and if he was lucky enough (or wise enough) to choose the one with the portrait, then he could claim Portia as his bride. On the lid of each casket was an inscription to help the suitor choose wisely.

Now, suppose Portia wished to choose her husband not on the basis of virtue, but simply on the basis of intelligence. She had the following inscriptions put on the caskets.

Gold	Silver	Lead
THE PORTRAIT IS IN THIS CASKET	THE PORTRAIT IS NOT IN THIS CASKET	THE PORTRAIT IS NOT IN THE GOLD CASKET

Portia explained to the suitor that of the three statements, at most one was true.

Which casket should the suitor choose?

67b.

Portia's suitor chose correctly, so they married and lived quite happily—at least for a while. Then, one day, Portia had the following thoughts: "Though my husband showed *some* intelligence in choosing the right casket, the problem wasn't really *that* difficult. Surely, I could have made the problem harder and gotten a *really* clever husband." So she forthwith divorced her husband and decided to get a cleverer one.

This time she had the following inscriptions put on the caskets:

Gold — THE PORTRAIT IS NOT IN THE SILVER CASKET

Silver — THE PORTRAIT IS NOT IN THIS CASKET

Lead — THE PORTRAIT IS IN THIS CASKET

Portia explained to the suitor that at least one of the three statements was true and that at least one of them was false.

Which casket contains the portrait?

Epilogue

As fate would have it, the first suitor turned out to be Portia's ex-husband. He was really quite bright enough to figure out this problem too. So they were remarried. The husband took Portia home, turned her over his knee, gave her a good sound spanking, and Portia never had any foolish ideas again.

B. THE SECOND TALE

Portia and her husband did, as a matter of fact, live happily ever after. They had a daughter Portia II—henceforth to be called "Portia." When the young Portia grew to young womanhood, she was both clever and beautiful, just like her mommy. She also decided to select her husband by the casket method. The suitor had to pass *two* tests in order to win her.

68a. The First Test.

In this test each lid contained *two* statements, and Portia explained that no lid contained more than one false statement.

Gold	Silver	Lead
(1) THE PORTRAIT IS NOT IN HERE	(1) THE PORTRAIT IS NOT IN THE GOLD CASKET	(1) THE PORTRAIT IS NOT IN HERE
(2) THE ARTIST OF THE PORTRAIT IS FROM VENICE	(2) THE ARTIST OF THE PORTRAIT IS REALLY FROM FLORENCE	(2) THE PORTRAIT IS REALLY IN THE SILVER CASKET

Which casket contains the portrait?

68b. The Second Test.

If the suitor passed the first test, he was taken into another room in which there were three more caskets. Again each casket had two sentences inscribed on the lid. Portia explained that on one of the lids, both statements were true; on another, both statements were false; and on the third, one statement was true and one was false.

Gold	Silver	Lead
(1) THE PORTRAIT IS NOT IN THIS CASKET	(1) THE PORTRAIT IS NOT IN THE GOLD CASKET	(1) THE PORTRAIT IS NOT IN THIS CASKET
(2) IT IS IN THE SILVER CASKET	(2) IT IS IN THE LEAD CASKET	(2) IT IS IN THE GOLD CASKET

Which casket contains the portrait?

C. INTRODUCING BELLINI AND CELLINI

The suitor of the last tale passed both tests and happily claimed Portia II as his bride. They lived happily ever after and had a lovely daughter Portia III—henceforth to be called "Portia." When she grew up to young womanhood, she was born smart and beautiful—just like her mommy and grandmommy. She also decided to choose her husband by the casket method. The suitor had to pass *three* tests in order to win her! The tests were quite ingenious. She went back to her grandmother's idea of having only one statement inscribed on each casket rather than two. But she introduced the following new wrinkle: She explained to the suitor that each casket was fashioned by one of two famous Florentine craftsmen—Cellini or Bellini. Whenever Cellini fashioned a casket, he always put a false inscription on it, whereas Bellini put only true inscriptions on his caskets.

69a. The First Test.

In this unusual test the suitor (if he guessed blindly) would have a two out of three rather than a one out of three chance. Instead of using a portrait, Portia used a dagger which was placed in one of the three caskets; the other two caskets were empty. If the suitor could *avoid the casket with the dagger,* then he could take the next test. The inscriptions on the caskets were as follows:

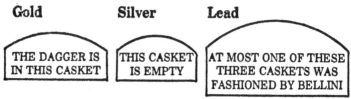

Gold — THE DAGGER IS IN THIS CASKET

Silver — THIS CASKET IS EMPTY

Lead — AT MOST ONE OF THESE THREE CASKETS WAS FASHIONED BY BELLINI

Which casket should the suitor choose?

69b. The Second Test. _____

In this test, the suitor's chances (if he guessed blindly) were one out of two. Portia used only two caskets, gold and silver, and one of them contained her portrait (no dagger was used in this test). Again each casket was fashioned either by Cellini or Bellini. The caskets read:

Gold

Silver

THE PORTRAIT IS NOT IN HERE

EXACTLY ONE OF THESE TWO CASKETS WAS FASHIONED BY BELLINI

Which casket should the suitor choose in order to find the portrait?

69c. The Third Test _____

If the suitor passed these two tests, he was led into another room containing a gold, silver, and lead casket. Again, each casket was fashioned by either Cellini or Bellini. Now in this test, the suitor's chances were one out of three (if he guessed blindly); Portia used a portrait of herself, and the portrait was in one of the caskets. To pass the test, the suitor had to (1) select the casket containing the portrait; (2) tell the maker of each casket.

The three inscriptions read:

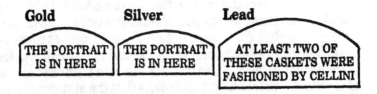

Gold

Silver

Lead

THE PORTRAIT IS IN HERE

THE PORTRAIT IS IN HERE

AT LEAST TWO OF THESE CASKETS WERE FASHIONED BY CELLINI

What is the solution?

70.

The fourth and final tale is the most baffling of all, and it illustrates a logical principle of basic importance.

The suitor of the last story passed all three tests and happily claimed Portia III as his bride. They had many children, great-grandchildren, etc.

Several generations later a descendant was born in America who looked so much like the ancestral portraits that she was named Portia Nth—henceforth to be referred to as "Portia." When this Portia grew to young womanhood she was both clever and beautiful—just like all the other Portias. In addition, she was highly vivacious and a bit on the mischievous side. She also decided to select her husband by the casket method (which was somewhat of an anomaly in modern New York, but let that pass).

The test she used appeared simple enough; she had only two caskets, silver and gold, in one of which was Portia's portrait. The lids bore the following inscriptions:

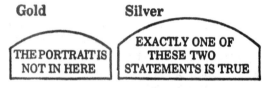

Gold

THE PORTRAIT IS
NOT IN HERE

Silver

EXACTLY ONE OF
THESE TWO
STATEMENTS IS TRUE

Which casket would you choose? Well, the suitor reasoned as follows. If the statement on the silver casket is true, then it is the case that exactly one of the two statements is true. This means that the statement on the gold casket must be false. On the other hand, suppose the statement on the silver casket is false. Then it is not the case that exactly one of the statements is true; this means that the statements are either both true or both false. They can't both be true (under the assumption that the second is false), hence they are both false. Therefore again, the statement on the gold casket is false. So regardless of whether the statement on the silver casket is true or false, the statement on the gold casket must be false. Therefore the portrait must be in the gold casket.

So the suitor triumphantly exclaimed, "The portrait must be in the gold casket" and opened the lid. To his utter horror the gold casket was empty! The suitor was stunned and claimed that Portia had deceived him. "I don't stoop to deceptions," laughed Portia, and with a haughty, triumphant, and disdainful air opened the silver casket. Sure enough, the portrait was there.

Now, what on earth went wrong with the suitor's reasoning?

"Well, well!" said Portia, evidently enjoying the situation enormously, "so your reason didn't do you much good, did it? However, you seem like a very attractive young man, so I think I'll give you another chance. I really shouldn't do this, but I will! In fact, I'll forget the last test and give you a simpler one in which your chances of winning me will be two out of three rather than one out of two. It resembles one of the tests given by my ancestor Portia III. Now *surely* you should be able to pass this one!"

So saying, she led the suitor into another room in which there were three caskets—gold, silver, and lead. Portia explained that one of them contained a dagger and the other two were empty. To win her, the suitor merely need choose one of the empty ones. The inscriptions on the caskets read as follows:

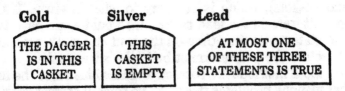

Gold	Silver	Lead
THE DAGGER IS IN THIS CASKET	THIS CASKET IS EMPTY	AT MOST ONE OF THESE THREE STATEMENTS IS TRUE

(Compare this problem with the first test of Portia III! Doesn't it seem to be exactly the same problem?)

Well, the suitor reasoned *very* carefully this time as follows: Suppose statement (3) is true. Then both other statements must be false—in particular (2) is false, so the dagger is then in the silver casket. On the other hand, if (3) is false, then there must be at least two true statements present, hence (1) must be one of them, so in this case the dagger is in the gold casket. In either case the lead casket is empty.

So the suitor chose the lead casket, opened the lid, and to his horror, there was the dagger! Laughingly, Portia opened the other two caskets and they were empty!

I'm sure the reader will be happy to hear that Portia married her suitor anyhow. (She had decided this *long* before the tests, and merely used the tests to tease him a little). But this still leaves unanswered the question: *What was wrong with the suitor's reasoning?*

SOLUTIONS

67a. _____

The statements on the gold and lead caskets say the opposite, hence one of them must be true. Since at most one of the three statements is true, then the statement on the silver casket is false, so the portrait is actually in the silver casket.

This problem could be alternatively solved by the following method: If the portrait were in the gold casket, we would have *two* true statements (namely on the gold and lead caskets), which is contrary to what is given. If the portrait were in the lead casket, we would again have two true statements (this time on the lead and silver caskets). Therefore the portrait must be in the silver casket.

Both methods are correct, and this illustrates the fact that in many problems there can be several correct ways of arriving at the same conclusion.

67b.

If the portrait were in the lead casket, then all three statements would be true, which is contrary to what is given. If the portrait were in the silver casket, then all three statements would be false, which is again contrary to what is given. Therefore the portrait must be in the gold casket (and we have the first two statements true and the third one false, which is consistent with what is given).

68a.

We can immediately rule out the lead casket, for if the portrait were there, then both statements on the lead casket would be false. So the portrait is in the gold or the silver casket. Now, the first statements on the gold and silver caskets agree, so they are both true or both false. If they are both false, the second statements are both true—but they cannot be both true since they are contradictory. Therefore the first statements are both true, so the portrait cannot be in the gold casket. This proves that the portrait is in the silver casket.

68b.

If the portrait is in the gold casket, then the gold and silver casket lids each contain two false statements. If it is in the silver casket, then the silver and lead caskets each contain one true and one false statement. Therefore the portrait is in the lead casket (and the silver casket lid contains both true statements; the lead, both false; and the gold, one true and one false).

69a.

Suppose the lead casket had been fashioned by Bellini. Then the statement would be true, hence the other caskets must have been fashioned by Cellini. This means that the other statements are both false—in particular the statement on the silver casket is false, so the dagger is in the silver casket. Thus, if the lead casket is the work of Bellini, then the silver casket contains the dagger.

Now, suppose the lead casket had been fashioned by Cellini. Then the statement is false, so at least two caskets were fashioned by Bellini. This means that both the gold and silver caskets are Bellini caskets (since the lead one is assumed Cellini). Then the statements on both the gold and silver are true. In particular, the one on the gold is true. So in this case, the dagger lies in the gold casket.

In neither case can the dagger be in the lead casket, so the suitor should choose the lead casket.

69b.

If the silver casket is a Bellini, then the statement is true, in which case the gold casket is a Cellini. Suppose the silver casket is a Cellini. Then it is not the case that exactly one of the caskets is a Bellini. This means that the gold is a Cellini (for if it were a Bellini, then it *would* be the case that exactly one is a Bellini!) Thus, whether the silver is Bellini or Cellini, the gold is surely a Cellini. Therefore the statement on the gold casket is false, so the portrait is in the gold casket.

69c.

We first show that the lead casket must be a Bellini. Suppose it were a Cellini. Then the statement is false, which means that there must be at least two Bellinis, which must be silver and gold. This is impossible, since the portrait can't be in both the silver and gold caskets. Thus the lead casket is really a Bellini. Hence the statement on it is true, so there are at least two Cellinis. This means that the gold and silver are both Cellinis. Hence the statements on both of them are false, so the portrait is neither in the gold nor the silver caskets. Therefore the portrait is in the lead casket.

Also, we have proved that the lead casket is a Bellini and the other two are Cellinis, which answers the second question.

70. _____

The suitor should have realized that without any information given about the truth or falsity of any of the sentences, nor any information given about the relation of their truth-values, the sentences could say anything, and the object (portrait or dagger, as the case may be) could be anywhere. Good heavens, I can take any number of caskets that I please and put an object in one of them and then write any inscriptions at all on the lids; these sentences won't convey any information whatsoever. So Portia was not really lying; all she said was that the object in question was in one of the boxes, and in each case it really was.

The situation would have been very different with any of the previous Portia stories, if the object had not been where the suitor figured it out to be; in this case one of the old Portias would have had to have made a false statement somewhere along the line (as we will soon see).

Another way to look at the matter is that the suitor's error was to assume that each of the statements was either true or false. Let us look more carefully at the first test of Portia Nth, using two caskets. The statement on the gold casket, "The portrait is not in here," is certainly either true or false, since either the portrait is in the gold casket or it isn't. It happened to be true, as a matter of fact, since Portia had actually placed the portrait in the silver casket. Now, given that Portia did put the portrait in the silver casket, was the statement on the silver casket true or false? It couldn't be either one without getting into a paradox! Suppose it were true. Then exactly one of the statements is true, but since the first statement (on the gold casket) is true, then this statement is false. So if it is true, it is false.

On the other hand, suppose this statement on the silver casket is false. Then the first is true, the second is false, which means that exactly one of the statements is true, which is what this statement asserts, hence it would have to be true! Thus either assumption, that the statement is true or is false, leads to a contradiction.

It will be instructive to compare this test with the second test given by Portia III, which also used just two caskets. The gold casket said the same thing as the gold of the problem, "The portrait is not in here," but the silver casket, instead of saying "Exactly one of these two statements is true," said "Exactly one of these two caskets was fashioned by Bellini." Now, the reader may wonder what significant difference there is between these two statements, given that Bellini inscribed only true statements and Cellini only false ones. Well, the difference, though subtle, is basic. The statement, "Exactly one of these two caskets was fashioned by Bellini" is a statement which *must* be true or false; it is a historic statement about the physical world—either it is or it is not the case that Bellini made exactly one of the two caskets. Suppose, in the Portia III problem, that the portrait had been found to be in the silver casket instead of the cold casket. What would you conclude: that the statement on the silver casket was neither true nor false? That would be the wrong conclusion! The statement, as I have pointed out, really is either true or false. The correct conclusion to draw is that if the portrait had been in the silver casket, then Portia III would have been lying in saying what she did about Bellini and Cellini. By contrast, the modern Portia could place the portrait in the silver casket without having lied, since she said nothing about the truth-values of the statements.

The whole question of the truth-values of statements which refer to their own truth-values is a subtle and basic aspect of modern logic and will be dealt with again in later chapters.

6. From the Files of Inspector Craig

A. FROM THE FILES OF INSPECTOR CRAIG

Inspector Leslie Craig of Scotland Yard has kindly consented to release some of his case histories for the benefit of those interested in the application of logic to the solution of crimes.

71.

We shall start with a simple case. An enormous amount of loot had been stolen from a store. The criminal (or criminals) took the heist away in a car. Three well-known criminals A,B,C were brought to Scotland Yard for questioning. The following facts were ascertained:

(1) No one other than A,B,C, was involved in the robbery.
(2) C never pulls a job without using A (and possibly others) as an accomplice.
(3) B does not know how to drive.

Is A innocent or guilty?

72.

Another simple case, again of robbery: A,B,C were brought in for questioning and the following facts were ascertained:

(1) No one other than A,B,C was involved.
(2) A never works without at least one accomplice.
(3) C is innocent.

Is B innocent or guilty?

73. The Case of the Identical Twins. _____

In this more interesting case, the robbery occurred in London. Three well-known criminals A,B,C were rounded up for questioning. Now, A and C happened to be identical twins and few people could tell them apart. All three suspects had elaborate records, and a good deal was known about their personalities and habits. In particular, the twins were quite timid, and neither one ever dared to pull a job without an accomplice. B, on the other hand, was quite bold and despised ever using an accomplice. Also several witnesses testified that at the time of the robbery, one of the two twins was seen drinking at a bar in Dover, but it was not known which twin.

Again, assuming that no one other than A,B,C was involved in the robbery, which ones are innocent and which ones guilty?

74. _____

"What do you make of these three facts?" asked Inspector Craig to Sergeant McPherson.

(1) If A is guilty and B is innocent, then C is guilty.
(2) C never works alone.
(3) A never works with C.
(4) No one other than A,B or C was involved, and at least one of them is guilty.

The Sergeant scratched his head and said, "Not much, I'm afraid, Sir. Can you infer from these facts which ones are innocent and which ones are guilty?"

"No," responded Craig, "but there is enough material here to definitely indict one of them."

Which one is necessarily guilty?

75. The Case of McGregor's Shop. _____

Mr. McGregor, a London shopkeeper, phoned Scotland Yard that his shop had been robbed. Three suspects A,B,C were rounded up for questioning. The following facts were established:

(1) Each of the men A,B,C had been in the shop on the day of the robbery, and no one else had been in the shop that day.
(2) If A was guilty, then he had exactly one accomplice.
(3) If B is innocent, so is C.
(4) If exactly two are guilty, then A is one of them.
(5) If C is innocent, so is B.

Whom did Inspector Craig indict?

76. Case of the Four. _____

This time four suspects A,B,C,D were rounded up for questioning concerning a robbery. It was known for sure that at least one of them was guilty and that no one outside these four was involved. The following facts turned up:

(1) A was definitely innocent.
(2) If B was guilty, then he had exactly one accomplice.
(3) If C was guilty, then he had exactly two accomplices.

Inspector Craig was especially interested in knowing whether D was innocent or guilty, since D was a particularly dangerous criminal. Fortunately, the above facts are sufficient to determine this. Is D guilty or not?

Inspector Craig frequently used to go to court to observe cases—even those in which he was not himself involved. He did this just as an exercise in logic—to see which cases he could figure out. Here are some of the cases he observed.

77.The Case of the Stupid Defense Attorney. _____
A man was being tried for participation in a robbery. The prosecutor and the defense attorney made the following statements:

> *Prosecutor* / If the defendant is guilty, then he had an accomplice.
> *Defense Attorney* / That's not true!

Why was this the worst thing the defense attorney could have said?

78. _____
This and the next case involve the trial of three men, A,B,C, for participation in a robbery.
　　In this case, the following two facts were established:

> (1) If A is innocent or B is guilty, then C is guilty.
> (2) If A is innocent, then C is innocent.

Can the guilt of any particular one of the three be established?

79. _____
In this case, the following facts were established:

> (1) At least one of the three is guilty.
> (2) If A is guilty and B is innocent, then C is guilty.

This evidence is insufficient to convict any of them, but it does point to two of them such that one of these two has to be guilty. Which two are they?

80. _____

In this more interesting case, four defendants A,B,C,D were involved and the following four facts were established:

(1) If both A and B are guilty, then C was an accomplice.
(2) If A is guilty, then at least one of B,C was an accomplice.
(3) If C is guilty, then D was an accomplice.
(4) If A is innocent then D is guilty.

Which ones are definitely guilty and which ones are doubtful?

81. _____

This case again involves four defendants, A,B,C,D. The following facts were established:

(1) If A is guilty, then B was an accomplice.
(2) If B is guilty then either C was an accomplice or A is innocent.
(3) If D is innocent then A is guilty and C is innocent.
(4) If D is guilty, so is A.

Which ones are innocent and which ones are guilty?

C. SIX EXOTIC CASES _____

82. Was It a Wise Thing to Say? _____

On a small island a man was being tried for a crime. Now, the court knew that the defendant was born and bred on the neighboring island of knights and knaves. (We recall that knights always tell the truth and knaves always lie.) The defendant was allowed to make only one statement in his own defense. He thought for a while and then came out with this statement: "The person who actually committed this crime is a knave."

Was this a wise thing for him to have said? Did it help or injure his case? Or did it make no difference?

83. The Case of the Uncertain Prosecutor. _____

On another occasion two men X,Y were being tried for a crime on this island. Now the most curious aspect of this case is that the prosecuting attorney was known to be either a knight or a knave. He made the following two statements in court:

(1) X is guilty.
(2) X and Y are not both guilty.

If you were on the jury, what would you make of this? Could you come to any conclusion about the guilt of either X or Y? What would be your opinion about the veracity of the prosecutor?

84. _____

In the above situation, suppose, instead, the prosecutor had made the following two statements:

(1) Either X or Y is guilty.
(2) X is not guilty.

What would you conclude?

85. _____

In the same situation, suppose, instead, the prosecutor had made the following two statements:

(1) Either X is innocent or Y is guilty.
(2) X is guilty.

What would you conclude?

86.

This case took place on the island of knights, knaves, and normals. We recall that knights always tell the truth, knaves always lie, and normals sometimes lie and sometimes tell the truth.

Three inhabitants of the island, A,B, and C, were being tried for a crime. It was known that the crime was committed by only one of them. It was also known that the one who committed the crime was a knight, and the only knight among them. The three defendants made the following statements:

A: I am innocent.
B: That is true.
C: B is not normal.

Which one is guilty?

87.

This, the most interesting case of all, bears a superficial resemblance to the above but is really quite different. It also took place on the island of knights, knaves, and normals.

The principal actors in this case were the defendant, the prosecutor, and the defense attorney. The first baffling thing was that it was known that one of them was a knight, one a knave, and one normal, though it was not known which was which. Even stranger, the court knew that if the defendant was not guilty, then the guilty one was either the defense attorney or the prosecutor. It was also known that the guilty one was not a knave. The three made the following statements in court:

Defendant / I am innocent.
Defense Attorney / My client is indeed innocent.
Prosecutor / Not true, the defendant is guilty.

These statements certainly seemed natural enough. The jury convened, but could not come to any decision; the above evidence was insufficient. Now, this island was a British possession at the time, hence the government wired to Scotland Yard asking whether they could send Inspector Craig to come over to help settle the case.

Several weeks later Inspector Craig arrived, and the trial was reconvened. Craig said to himself, "I want to get to the bottom of this!" He wanted to know not only who was guilty, but also which one was the knight, which the knave, and which the normal. So he decided to ask just enough questions to settle these facts. First he asked the prosecutor, "Are you, by any chance, the guilty one?" The prosecutor answered. Inspector Craig thought for a while, and then he asked the defendant, "Is the prosecutor guilty?" The defendant answered, and Inspector Craig knew everything.

Who was guilty, who was normal, who was the knight, and who was the knave?

SOLUTIONS

71. _____

I shall first show that at least one of A,C is guilty. If B is innocent, then it's obvious that A and/or C is guilty—since by (1), no one other than A,B,C is guilty. If B is guilty, then he must have had an accomplice (since he can't drive), so again A or C must be guilty. So A or C (or both) are guilty. If C is innocent, then A must be a guilty one. On the other hand, if C is guilty, then by statement (2), A is also guilty. Therefore A is guilty.

72. _____

This is even simpler. If A is innocent, then, since C is innocent, B must be guilty—by (1). If A is guilty, then, by (2), he had an accomplice, who couldn't be C—by (3), hence must be B. So in either case, B is guilty.

73. _____

Suppose B were innocent. Then one of the twins must be guilty. This twin must have had an accomplice who couldn't be B hence must have been the other twin. But this is impossible since one of the twins was in Dover at the time. Therefore B is guilty. And since B always works alone, both twins are innocent.

74. _____

B must be guilty. This can be shown by either of the following arguments.

Argument One: Suppose B were innocent. Then if A were guilty, C would also be guilty—by statement (1)—but this would mean that A worked with C, which contradicts statement (3). Therefore A must be innocent. Then C is the only guilty one, contradicting statement (2). Therefore B is guilty.

Argument Two: A more direct argument is this: (a) Suppose A is guilty. Then by (1), B and C cannot both be innocent, hence A must have had an accomplice. This accomplice couldn't have been C—by (3), hence must have been B. So if A is guilty, B is also guilty. (b) Suppose C is guilty. Then he had an accomplice—by (2)—which couldn't be A—by (3)—hence must again be B.
(c) If neither A nor C is guilty, then B certainly is!

75.

Inspector Craig indicted Mr. McGregor for falsely claiming there was a robbery, when in fact there couldn't have been one! His reasoning was as follows.

Step One: Suppose A were guilty. Then he had exactly one accomplice—by (2). Then one of B,C is guilty and the other innocent. This contradicts (3) and (5), which jointly imply that B,C are either both innocent or both guilty. Therefore A must be innocent.

Step Two: Again, by (3) and (5), B and C are both guilty or both innocent. If they were both guilty, then they were the only guilty ones (since A is innocent). Then there would be exactly two guilty ones, which by statement (4) would imply that A is guilty. This is a contradiction, since A is innocent. Therefore B,C are both innocent.

Step Three: Now it is established that A,B,C are all innocent. Yet, by statement (1), no one other than A,B,C had been in the shop on the day of the robbery and could have commited the robbery. Ergo, there was no robbery and McGregor was lying.

Epilogue:

Confronted by Craig's irrefutable logic, McGregor broke down and confessed that he had indeed lied and was trying to collect insurance.

76.

If B was guilty, then by (2) exactly two people were involved; if C was guilty, then by (3) exactly three people were involved. These can't both be the case, hence at least one of B,C is innocent. A is also innocent, so there are at most two guilty ones. Therefore C did not have exactly two accomplices, so by (3) C must be innocent. If B is guilty then he had exactly one accomplice, who must have been D (since A,C are both innocent). If B is innocent, then A,B,C are all innocent, in which case D must be guilty. So regardless of whether B is guilty or innocent, D must be guilty. Therefore D is guilty.

77.

The prosecutor said, in effect, that the defendant didn't commit the crime alone. The defense attorney denied this, which is tantamount to saying that the defendant *did* commit the crime alone.

78.

This is extremely simple. By (1), if A is innocent, then C is guilty (because if A is innocent then the statement, "either A is innocent or B is guilty" is true). By (2), if A is innocent then C is innocent. Therefore if A is innocent, then C is both guilty and innocent, which is impossible. Therefore A must be guilty.

79.

The two are B and C; at least one of them must be guilty. For, suppose A is innocent. Then B or C must be guilty by (1). On the other hand suppose A is guilty. If B is guilty, then certainly at least one of B,C is guilty. But suppose that B is innocent. Then A is guilty and B is innocent, hence by (2), C must be guilty, so again either B or C is guilty.

80.

We first show that if A is guilty, so is C. Well, suppose A is guilty. Then by (2), either B or C is guilty. If B is innocent, then it must be C who is guilty. But suppose B is guilty. Then A and B are both guilty, hence by (1) C is guilty too. This proves that if A is guilty, so is C. Also, by (3), if C is guilty so is D. Combining these two facts, we see that if A is guilty, so is D. But by (4), if A is innocent, so is D. Therefore, regardless of whether A is guilty or innocent, D must be guilty. So D is definitely guilty. The rest are all doubtful.

81.

The answer is that all of them are guilty. By (3), if D is innocent then A is guilty. By (4), if D is guilty, then A is guilty. So whether D is innocent or guilty, A must be guilty. Hence by (1), B is also guilty. Hence by (2), either C is guilty or A is innocent. But we already know that A is not innocent, therefore C must be guilty. Finally, by (3), if D is innocent then C is innocent. But we have proved that C is not innocent, hence D must be guilty. So all of them are guilty.

82.

Yes, it was wise; it acquitted him. For suppose the defendant is a knight. Then his statement is true, hence the guilty man is a knave, hence the defendant must be innocent. On the other hand, suppose the defendant is a knave. Then his statement is false, so the criminal is actually a knight, so again the defendant is innocent.

83.

Suppose the prosecutor were a knave. Then (1) and (2) would both be false. Since (1) is false, then X is innocent. Since (2) is false, then X,Y are both guilty—hence X is guilty. This is a contradiction. So the prosecutor must be a knight. Hence X really is guilty, and since they are not both guilty, Y must be innocent. Therefore X is guilty, Y is innocent, and the prosecutor is a knight.

84.

If the prosecutor were a knave, then it would be the case that (1) X and Y are both innocent; (2) X is guilty. Again, this is a contradiction, so the prosecutor is a knight, X is innocent, and Y is guilty.

85.

Again, suppose the prosecutor were a knave. Then (1) is false, so X is guilty and Y is innocent. Hence X is guilty. But (2) is also false, hence X is innocent: another contradiction. Hence the prosecutor is again a knight. Therefore, by (2), X is guilty. Then by (1) (since X is not innocent), Y must be guilty. So this time X and Y are both guilty.

86.

A cannot be a knight, for if he were he would be guilty and wouldn't have lied about being innocent. Also A cannot be a knave, for if he were, his statement would be false, hence he would be guilty and hence would be a knight. Therefore A is normal, hence also innocent. Since A is innocent, B's statement is true. Therefore B is not a knave; he is a knight or normal. Suppose B were normal. Then C's statement would be false, hence C would be a knave or a normal. This would mean that none of A,B,C is a knight, hence none of them is guilty, contrary to what is given. Therefore B cannot be normal, he must be a knight and hence guilty.

87.

Before Craig Arrived: To begin with,[1] A cannot be a knave, because if he were a knave his statement would be false, hence he would be guilty, contrary to the given condition that the knave is not guilty. Therefore A is either a knight or normal.

Possibility One: A is a knight: Then his statement is true, hence he is innocent. Then B's statement is also true, hence B is a knight or normal. But A is the knight, so B is normal.

[1] We are letting A be the defendant, B the defense attorney, and C the prosecutor.

This leaves C as the knave. So, since it is known that the knave is not guilty, B is guilty.

Possibility Two: A is normal and innocent: Then B's statement is again true, hence B is the knight (since A is the normal one). So, since A is innocent, and C, being the knave, is innocent, then B is guilty.

Possibility Three: A is normal and guilty: Then the prosecutor's statement was true, so the prosecutor must be a knight (again, he can't be normal, since A is). This leaves B as the knave.

Let us summarize the three possibilities:

	(1)	(2)	(3)
Defendant	Innocent Knight	Innocent Normal	Guilty Normal
Defense Attorney	Guilty Normal	Guilty Knight	Innocent Knave
Prosecutor	Innocent Knave	Innocent Knave	Innocent Knight

All three possibilities are consistent with the statements made before Craig arrived.

After Craig Arrived: Craig asked the prosecutor whether he was guilty. Now, he already knew that he was innocent (because in all of the above three possibilities, the prosecutor is innocent); so the prosecutor's answer would only serve to let Craig know whether the prosecutor was a knight or a knave. Had he truthfully answered "No," revealing himself to be a knight, then Craig would have known that possibility (3) was in fact the only one, hence he would not have asked any more questions. But after the prosecutor's answer, Craig *did* ask more questions. Therefore the prosecutor must have been a knave and answered "Yes." So now Craig (as well as the reader) knows that possibility (3) is out, which leaves (1) and

(2). This means that the defense attorney is in fact the guilty one, but it is still unknown which of the defendant and the defense attorney is the knight and which is normal. Craig then asked the defendant whether the prosecutor was guilty, and after he was answered, he knew the entire situation. Well, a knight would have to answer "No" to this question, whereas a normal could answer it either "Yes" or "No." Had the answer been "No," there would have been no way of Craig's knowing whether the defendant was a knight or a normal. But Craig did know, therefore he must have gotten a "Yes" answer. Therefore the defendant is normal and the defense attorney is a knight (though guilty).

7. How to Avoid Werewolves— and Other Practical Bits of Advice

This chapter is concerned more with the practical than the recreational aspects of logic. There are many situations in life in which it is good to have one's wits about one. So I shall now give you detailed, step-by-step instructions which will teach you: (A) how to avoid werewolves in the forest; (B) how to choose a bride; (C) how to defend yourself in court; (D) how to marry a king's daughter.

Of course, I cannot absolutely guarantee that you will actually meet with any of these situations, but as the White Knight wisely explained to Alice, it is well to be provided for *everything*.

A. WHAT TO DO IN THE FOREST OF WEREWOLVES

Suppose you are visiting a forest in which every inhabitant is either a knight or a knave. (We recall that knights always tell the truth and knaves always lie.) In addition, some of the inhabitants are werewolves and have the annoying habit of sometimes turning into wolves at night and devouring people. A werewolf can be either a knight or a knave.

88.

You are interviewing three inhabitants, A, B, and C, and it is known that exactly one of them is a werewolf. They make the following statements:

> A: C is a werewolf.
> B: I am not a werewolf.
> C: At least two of us are knaves.

Our problem has two parts:

> (a) Is the werewolf a knight or a knave?
> (b) If you have to take one of them as a traveling companion, and it is more important that he not be a werewolf than that he not be a knave, which one would you pick?

89.

Again, each of A, B, C is a knight or a knave and exactly one of them is a werewolf. They make the following statements:

> A: I am a werewolf.
> B: I am a werewolf.
> C: At most one of us is a knight.

Give a complete classification of A, B, and C.

90.

In this and the next two problems there are again three inhabitants A, B, C, each of whom is either a knight or a knave. However only two of them, A, B, make statements. But in these statements, the word "us" refers to the three people A, B, C—not to just A and B.

Suppose A, B make the following statements:

> A: At least one of the three of us is a knight.
> B: At least one of the three of us is a knave.

Given that at least one of them is a werewolf, and that none of them is both a knight and a werewolf, which ones are werewolves?

91. _____

This time, we get the following statements:

> A: At least one of the three of us is a knave.
> B: C is a knight.

Given that there is exactly one werewolf and that he is a knight, who is the werewolf?

92. _____

In this problem we get the following two statements:

> A: At least one of the three of us is a knave.
> B: C is a werewolf.

Again, there is exactly one werewolf and he is a knight. Who is he?

93. _____

In this problem we are given that there is exactly one werewolf and that he is a knight, and that the other two are knaves. Only one of them, B, makes a statement: "C is a werewolf."
Who is the werewolf?

94. _____

Here is an elegantly simple one involving just two inhabitants, A and B. Just one of them is a werewolf. They make the following statements:

A: The werewolf is a knight.
B: The werewolf is a knave.

Which one would you select for your traveling companion?

B. HOW TO WIN OR CHOOSE A BRIDE

95. How Do You Convince Her? _____

Suppose you are an inhabitant of the island of knights and knaves. You fall in love with a girl there and wish to marry her. However, this girl has strange tastes; for some odd reason she does not wish to marry a knight; she wants to marry only a knave. But she wants a rich knave, not a poor one. (We assume for convenience that everyone there is classified as either rich or poor.) Suppose, in fact, that you are a rich knave. You are allowed to make only one statement to her. How, in only one statement, can you convince her that you are a rich knave?

96. _____

Suppose, instead, the girl you love wants to marry only a rich knight. How, in one statement, could you convince her that you are a rich knight?

97. How to Choose a Bride. _____

This time you are a visitor to the island of knights and knaves. Every female there is either a knight or a knave. You fall in love with one of the females there—a girl named Elizabeth—and are thinking of marrying her. However, you want to know just what you are getting into; you do not wish to marry a knave. If you were allowed to question her, there would be no problem, but an ancient taboo of the island forbids a man to hold speech with any female unless he is already married to her. However, Elizabeth has a brother

Arthur who is also a knight or a knave (but not necessarily the same as his sister). You are allowed to ask just one question of the brother, but the question must be answerable by "Yes" or "No."

The problem is for you to design a question such that upon hearing the answer, you will know for sure whether Elizabeth is a knight or a knave. What question would you ask?

98. How to Choose a Bride on the Island of Bahava.

This time you are visiting the island of Bahava, in which there are knights, who always tell the truth, knaves, who always lie, and normals, who sometimes lie and sometimes tell the truth. Bahava, we recall, is a female liberationist island, hence the females are also called knights, knaves, or normals. Since you are an outsider, you are not subject to the injunction that a knight may marry only a knight and a knave only a knave, so you are free to marry the female of your choice.

Now, you are to pick a bride from among three sisters A,B,C. It is known that one of them is a knight, one a knave, and the other normal. But it is also known (to your horror!) that the normal one is a werewolf, but the other two are not. Now, let us assume that you don't mind marrying a knave (or a knight), but marrying a werewolf is going just a bit too far! You are allowed to ask any one question of your choice to any of the three sisters of your choice, but again the question must have a "Yes" or "No" answer.

What question would you ask?

C. YES, YOU ARE INNOCENT, BUT CAN YOU PROVE IT?

We now come to a particularly enticing group of puzzles. They all take place on the island of knights, knaves, and normals. You yourself are now one of the inhabitants of the island.

A crime has been committed on the island, and for some strange reason it is suspected that you are the criminal. You are brought to court and tried. You are allowed to make only one statement in your own behalf. Your purpose is to convince the jury that you are innocent.

99. _____

Suppose it is known that the criminal is a knave. Suppose also that you are a knave (though the court doesn't know this) but that you are nevertheless innocent of this crime. You are allowed to make only one statement. Your purpose is not to convince the jury that you are not a knave, but only that you are innocent of the crime. What would you say?

100. _____

Suppose you are in the same situation except for the fact that you are guilty. What statement could you make which would convince the jury (assuming they were rational beings) that you are innocent?

101. _____

In this problem, suppose it is known that the criminal is a knight. (This is no contradiction; a person doesn't necessarily have to lie in order to commit a crime.) Suppose also that you are a knight (but the jury doesn't know this) but innocent of the crime. What statement would you make?

102. _____

Here is a more difficult one. Suppose that in this problem it is known that the criminal is not normal—he is a knight or a knave. You yourself are innocent. What statement could you make which could be made by either a knight, a knave, or a normal in your position, which would convince the jury that you are innocent?

103. _____

Here is a much easier one. Again it is known that the criminal is not normal. Again you are not the criminal, but you are normal. What statement could you make which neither an innocent knight or knave could make which would convince the jury that you are innocent?

104. _____

Here is a more interesting one. Again it is known that the criminal is not normal. Let us suppose that (1) You are innocent; (2) you are not a knave.

Is there one single statement you could make which would simultaneously convince the jury of both of these facts?

105. _____

A sort of "dual" to the above problem is this: Suppose that again the guilty one is not normal and that you are an innocent but not a knight. Suppose that for some odd reason, you don't mind getting the reputation of being a knave or a normal, but you despise knights. Could you in one statement convince the jury that you are innocent but not a knight?

D. HOW TO MARRY A KING'S DAUGHTER

And now we come to the topic which I am sure you have all been anxiously waiting for!

106. _____

You are an inhabitant of the island of knights, knaves, and normals. You are in love with the King's daughter Margozita and wish to marry her. Now, the King does not wish his daughter to marry a normal. He says to her: "My dear, you really shouldn't marry a normal, you know. Normals are capricious, random, and totally unreliable. With a normal, you never know where you stand; one day he is telling you the truth, and the next day he is lying to you. What good is that? Now, a knight is thoroughly reliable, and with him you always know where you stand. A knave is really as good, because whenever he says anything, all you have to do is believe the opposite, so you still know how matters really are. Besides, I believe a man should stick to his principles. If a man believes in telling the truth, then let him always tell the truth. If he believes in lying, let him at least be consistent about it. But these wishy-washy bourgeois normals —no my dear, they are not for you!"

Well now, suppose that you are in fact not normal, so you have a chance. However, you must convince the King that you are not normal, otherwise he won't let you marry his daughter. You are allowed an audience with the King and you are allowed to make as many statements to him as you like. This problem has two parts.

(a) What is the smallest number of true statements you can make which will convince the King that you are not normal?

(b) What is the smallest number of false statements you can make which will convince the King that you are not normal?

107.

On another island of knights, knaves, and normals, the King has the opposite philosophy. He says to his daughter: "Darling, I don't want you to marry a knight or a knave; I want you to marry a good solid normal. You don't want to marry a knight, because knights are too sanctimonious. You don't want to marry a knave, because knaves are too treacherous. No, my dear, a good, conventional, bourgeois normal is just the thing for you!"

Suppose you are a normal on this island. Your job is to convince the King that you are normal.

    (a) What is the smallest number of true statements you could make which would convince the King that you are normal?

    (b) What is the smallest number of false statements you could make which would convince the King that you are normal?

108.

Here's a more difficult version of the above problem. The solution of this one constitutes an alternative (though unnecessarily complicated) solution of the preceding one, but the solution given for the previous one will not suffice to solve this one.

Again, you are a normal on an island of knights, knaves, and normals. Again the King wants his daughter to marry only a normal, but he also requires proof of exceptional ingenuity and intelligence. Therefore, to win his daughter, you must make a single statement in his presence which will simultaneously satisfy the following two requirements:

    (1) It must convince the King that you are normal.
    (2) It must make it impossible for the King to know whether the statement is true or false.

How can this be done?

SOLUTIONS

88.

C is either a knight or a knave. Suppose he is a knight. Then there really are at least two knaves, hence they must be A and B. Then B must be a werewolf (since he says he isn't but he is a knave). So if C is a knight, then the werewolf is a knave (since he must be B). On the other hand, suppose C is a knave. Then it is not true that at least two of them are knaves, so there is at most one knave. This knave must be C, hence A,B are both knights. Since A is a knight and claims that C is a werewolf, then C really is a werewolf. So in this case, the werewolf is again a knave—namely, he is C.

Therefore, regardless of whether C is a knight or a knave, the werewolf is a knave (though a different person in each case). So the answer to the first question is that the werewolf is a knave. Also, we have proved that the werewolf is either B or C; hence if you wish to choose someone who is definitely not a werewolf, then pick A.

89.

We first show that C is a knight. Suppose he were a knave. Then his statement would be false, hence there would be at least two knights. Then A,B would both have to be knights (since C is assumed a knave), which would mean that their statements were true and they were both werewolves, which contradicts the given conditions of the problem. Therefore C is a knight. Then there really are two knaves; these must be A and B. Then, since their statements are false, neither A nor B is a werewolf, so the werewolf must be C. Thus C is a knight and a werewolf; A and B are knaves and neither one a werewolf.

90.

If B were a knave then there would indeed be at least one knave among them, hence his statement would be true, but knaves don't make true statements. Therefore B is a knight. Then A's statement is true, so A is also a knight. So A and B are both knights. Since B is a knight, his statement is true so there is at least one knave. This knave must be C. Hence C is the one and only werewolf.

91.

A must be a knight for the same reasons that B was a knight in the last problem, namely that if A were a knave, it would be true that at least one of the three was a knave, and we would have a knave making a true statement. Since A is a knight, his statement is true, so there really is at least one knave present. If B were a knight, then C would also be (because of B's statement) and we would have three knights. But A says truthfully that there is at least one knave. Therefore B must be a knave. And since B says that C is a knight, C is really a knave. Thus A is the only knight, hence A is the werewolf.

92.

Again, because of A's statement, A must be a knight and there must be at least one knave. If B were a knight then C would be a werewolf, hence also a knight, and we would have three knights. Therefore B is a knave. Hence C is not a werewolf. Also B can't be a werewolf (since we are given that the werewolf is a knight). So again A is the werewolf.

93.

If B were a knight, then C would be a werewolf and also a knight and we would have two knights. So B is a knave. Hence C is not a werewolf. Also B, being a knave, is not a werewolf. So again A is the werewolf.

94.

You should select B. Suppose B is a knight. Then his statement is true, hence the werewolf is a knave, hence cannot be B. Suppose B is a knave. Then his statement is false, which means that the werewolf is actually a knight, hence again cannot be B.

95.

All you have to say is, "I am a poor knave." She will immediately know that you can't be a knight (since a knight would never lie and say he is a poor knave), hence you must be a knave. Hence also your statement is false, so you are not a poor knave. But you are a knave, hence you must be a rich knave.

96.

You say, "I am not a poor knight." She would reason that if you were a knave, you would indeed not be a poor knight, hence your statement would be true, hence you—a knave—would be making a true statement. Therefore you are a knight. Hence also your statement is true, so you are not a poor knight. But you are a knight, hence you must be a rich knight.

97.

This problem has several solutions. The simplest I can think of is that you ask, "Are you and Elizabeth of the same type?" The interesting thing is that if he answers "Yes," then Elizabeth must be a knight, regardless of whether the brother is a knight or knave, and if the brother answers "No," then Elizabeth must be a knave, regardless of what the brother is. Let us prove this.

Suppose he answers "Yes." Now, the brother is either a knight or a knave. If he is a knight, then his statement that Elizabeth is of the same type is true, hence Elizabeth must also be a knight. If he is a knave, then his statement is false, hence he and Elizabeth are of different types, which means that Elizabeth is again a knight. Thus if Arthur answers "Yes," Elizabeth is a knight.

Suppose Arthur answers "No." If he is a knight then he is telling the truth, hence he and Elizabeth are of different types, hence Elizabeth must be a knave. If he is a knave, then his statement is false, hence Elizabeth really is of the same type, hence must again be a knave. So if he answers "No," then Elizabeth is a knave.

98. _____

Again, there are several ways to solve this. The simplest and most elegant solution I know is to pick one of them—say A—and ask her, "Is B of lower rank than C?"[1]

Suppose A answers "Yes." Then you should pick B for your bride for the following reasons: Suppose A is a knight. Then B really is of lower rank than C, hence B is a knave and C is normal. In this case, B is not the werewolf (since C is). Suppose that A is a knave. Then B is actually of higher rank than C, which means that B is a knight and C normal, so again B is not a werewolf. If A is normal, then B is certainly not the werewolf, since A is. Thus, regardless of whether A is a knight, a knave, or a normal, if A answers "Yes" to your question, then you should pick B for your bride.

If A should answer "No," then it is the same as if she should assert that C is of lower rank that B, rather than that B is of lower rank than C, so in this case pick C for your bride.

99. _____

One statement which would acquit you is, "I am guilty." You, a knave, can actually say that, since it is false, and it will indeed acquit you, for the jury will correctly reason thus: If you really were guilty, then you would be a knave (since it is given that the criminal is known to be a knave), but then you, a knave, would be making a true statement. Thus the assumption that you are guilty leads to a contradiction, hence you are innocent.

[1] We recall that knights are of the highest rank, normals are of the middle rank, and knaves are of the lowest rank.

The above reasoning is an example of a *reductio ad absurdum* argument (proof of the falsity of a statement by reducing it to absurdity). A more direct argument the jury could have used is this: Either you are a knave or you are not (remember, the jury doesn't know whether or not you are a knave). If you are a knave then your statement is false, hence you are innocent. If you are not a knave, then you are certainly innocent, since the guilty one is a knave.

100. _____

No such statement is possible. If, after making a statement, the jury could rationally deduce that you are innocent, then, since they are rational and have used correct reasoning, it must be the case that you really are innocent. But this is contrary to the assumption that you are guilty.

101. _____

This is a sort of "dual" to problem 99, and, if anything, even simpler. All you need say is, "I am innocent." The jury will reason that if you are a knight (which they don't know) then your statement is true, hence you are innocent, and if you are not a knight, then again you are innocent, since the guilty one is known to be a knight.

102. _____

One solution is to say: "Either I am a knight and innocent, or I am a knave and guilty." Let's say you phrase it a bit more simply thus: "I am either an innocent knight or a guilty knave." The jury would then reason about you as follows:

Step One: Suppose he is a knight. Then his statement is true, hence he is either an innocent knight or a guilty knave. He can't be a guilty knave, since he is not a knave, hence he is an innocent knight. Hence he is innocent.

Step Two: Suppose he is a knave. Then his statement is false, hence he is neither an innocent knight nor a guilty knave. In particular, he is not a guilty knave. But he is a knave. Then he must be an innocent knave, hence innocent.

Step Three: If he is normal, then he is certainly innocent, since the guilty one is not normal.

103.

This is indeed quite simple. All you need to say is, "I am a knave." Neither a knight nor a knave could say that, hence you must be normal, hence also innocent.

104.

Yes, you could say, "I am not a guilty knight." The jury would reason this way:

Step One: Suppose he (meaning "you") were a knave. Then he is not a knight, hence certainly not a guilty knight, so his statement would be true. This is impossible, since knaves don't make true statements. Therefore he cannot be a knave.

Step Two: Now we know that he is either a knight or normal. If he is normal, he is innocent. Suppose he is a knight. Then his statement is true. Therefore he is not a guilty knight. But he is a knight. Hence he must be an innocent knight.

I might remark that you could alternatively have said, "Either I am not a knight or I am innocent," or you could have said, "If I am a knight then I am innocent."

105.

Yes, you could say, "I am a guilty knave." The jury would reason this way: "Obviously he is not a knight. So he is normal or a knave. If he is normal, he is innocent. Suppose he is a knave. Then his statement is false, so he is not a guilty knave. Hence he is an innocent knave."

106.

No amount of statements could possibly do this. Given any set of statements you make, a normal person could make the same statements, since a normal person can say anything. So there is no way you can marry this King's daughter. Sorry! Better luck on the next island!

107.

In both cases, one statement is enough. A true statement which would convince the King is: "I am not a knight." (Neither a knight nor a knave could say this.) A false statement which would do the job is: "I am a knave."

I wish to remark (in connection with the next problem) that if you make the first statement, then the King will know that although you are normal, you have just made a true statement; and if you make the second statement, the King will know that although you are normal, you have just made a false statement.

108.

Take any proposition whose truth or falsity is unknown to the King—for example, that you are now carrying exactly eleven dollars in your pocket. Then a statement you could make is: "Either I am normal and am now carrying exactly eleven dollars in my pocket, or else I am a knave."

A knave could never make that statement (because it is true that a knave is either a normal who is carrying eleven dollars or a knave). A knight also couldn't make that statement (because a knight is neither a normal who is carrying eleven dollars nor a knave). Therefore the King will know that you are normal, but he cannot know whether your statement is true or false without knowing how much money you are carrying.

8. Logic Puzzles

PREAMBLE

Many of the puzzles in this chapter deal with so-called *conditional* statements: statements of the form "If P is true then Q is true," where P, Q are statements under consideration. Before turning to puzzles of this type, we must carefully clear up some ambiguities which might arise. There are certain facts about such statements which everyone agrees on, but there are others about which there appears to be considerable disagreement.

Let us turn to a concrete example. Consider the following statement:

(1) If John is guilty, then his wife is guilty.

Everyone will agree that if John is guilty and if statement (1) is true, then his wife is also guilty.

Everyone will also agree that if John is guilty and his wife is innocent, then statement (1) must be false.

Now, suppose it is known that his wife is guilty, but it is not known whether John is guilty or innocent. Would you then say that statement (1) is true or not? Would you not say that whether John is guilty or whether he is innocent, his wife is guilty in any case? Or would you not say: If John is guilty then his wife is guilty, and if John is innocent then his wife is guilty?

Illustrations of this use of language abound in the literature: In Rudyard Kipling's story *Riki-Tiki-Tavi*, the cobra says to the terrified family, "If you move I will strike, and if you don't move I will strike." This means nothing more or less than: "I will strike." There is also the story of the Zen-master Tokusan, who used to answer all questions, as well as nonquestions, with blows from his stick. His famous saying is: "Thirty blows when you have something to say; thirty blows just the same when you have nothing to say."

The upshot is that if a statement Q is true outright, then so is the statement, "If P then Q" (as well as the statement, "If not P, then Q").

The most controversial case of all is this: Supposing P, Q are both false. Then is the statement, "If P then Q" true or false? Or does it depend on what P and Q are? Returning to our example, if John and his wife are both innocent, then should statement (1) be called true or not? We shall return to this vital question shortly.

A related question is this: We have already agreed that if John is guilty and his wife innocent, then statement (1) must be false. Is the converse true? That is, if statement (1) is false, does it follow that John must be guilty and his wife innocent? Put otherwise, is it the case that the *only* way that (1) can be false is that John be guilty and his wife innocent? Well, according to the way most logicians, mathematicians, and scientists use the words "if . . . then," the answer is "yes," and this is the convention we shall adopt. In other words, given any two statements P and Q, whenever I write "If P then Q" I shall mean nothing more nor less than "It is not the case that P is true and Q is false." In particular, this means that if John and his wife are both innocent, then statement (1) is to be regarded as true. For the only way the statement can be false is that John is guilty and his wife is innocent, and this state of affairs can't hold if John and his wife are both innocent. Stated otherwise, if John and his wife are both innocent, then it is certainly not the case that John is guilty and his wife is innocent, therefore the statement cannot be false.

The following is an even more bizarre example:

(2) If Confucius was born in Texas, then I am Dracula.

All statement (2) is intended to mean is that it is not the case that Confucius was born in Texas and that I am not Dracula. This indeed is so, since Confucius was not born in Texas. Therefore statement (2) is to be regarded as true.

Another way to look at the matter is that the only way (2) can be false is if Confucius was born in Texas and I am not Dracula. Well, since Confucius was not born in Texas, then it can't be the case that Confucius was born in Texas *and* that I am not Dracula. In other words, (2) cannot be false, so it must be true.

Now let us consider two arbitrary statements P, Q, and the following statement formed from them:

(3) If P then Q.

This statement is symbolized: $P \rightarrow Q$, and is alternatively read: "P implies Q." The use of the word "implies" may be somewhat unfortunate, but it has found its way into the literature in this sense. All the statement means, as we have seen, is that it is not the case that P is true and Q is false. Thus we have the following facts:

Fact 1: If P is false, then $P \rightarrow Q$ is automatically true.

Fact 2: If Q is true, then $P \rightarrow Q$ is automatically true.

Fact 3: The one and only way that $P \rightarrow Q$ can be false is that P is true and Q is false.

Fact 1 is sometimes paraphrased: "A false proposition implies any proposition." This statement came as quite a shock to many a philosopher (see Chapter 14, number 244, for a further discussion). Fact 2 is sometimes paraphrased: "A true proposition is implied by any proposition."

A Truth-Table Summary

Given any two statements P, Q, there are always exactly four possibilities: (1) P, Q are both true; (2) P is true and Q is false; (3) P is false and Q is true; (4) P, Q are both false.

One and only one of these possibilities must hold. Now, let us consider the statement, "If P then Q (symbolized: $P \rightarrow Q$). Can it be determined in which of the four cases it holds and in which ones it doesn't? Yes it can, by the following analysis:

Case 1: P *and* Q *are both true.* In this case Q is true, hence $P \rightarrow Q$ is true by Fact 2.

Case 2: P *is true and* Q *is false.* In this case, $P \rightarrow Q$ is false by Fact 3.

Case 3: P *is false and* Q *is true.* Then $P \rightarrow Q$ is true by Fact 1 (also by Fact 2).

Case 4: P *is false and* Q *is false.* Then $P \rightarrow Q$ is true by Fact 1.

These four cases are all summarized in the following table, called the *truth-table for implication.*

	P	Q	$P \rightarrow Q$
(1)	T	T	T
(2)	T	F	F
(3)	F	T	T
(4)	F	F	T

The first row, T,T,T (true, true, true), means that when P is true and Q is true, $P \rightarrow Q$ is true. The second row, T,F,F, means that when P is true and Q is false then $P \rightarrow Q$ is false. The third row says that when P is false and Q is true, $P \rightarrow Q$ is true, and the fourth row says that when P is false and Q is false, then $P \rightarrow Q$ is true.

We note that $P \rightarrow Q$ is true in three out of four of those cases; only in the second is it false.

Another Property of Implication. Another important property of implication is this: To show that a statement "If *P* then *Q*" holds, it suffices to assume *P* as premise and then show that *Q* must follow. In other words, if the assumption of *P* leads to *Q* as a conclusion, then the statement "If *P* then *Q*" is established.

We shall henceforth refer to this fact as *Fact 4.*

A. APPLICATION TO KNIGHTS AND KNAVES

109.

We have two people A,B, each of whom is either a knight or a knave. Suppose A makes the following statement: "If I am a knight, then so is B."

Can it be determined what A and B are?

110.

Someone asks A, "Are you a knight?" He replies, "If I'm a knight, then I'll eat my hat!"

Prove that A has to eat his hat.

111.

A says, "If I'm a knight, then two plus two equals four." Is A a knight or a knave?

112.

A says, "If I'm a knight, then two plus two equals five." What would you conclude?

113.

Given two people, A,B, both of whom are knights or knaves. A says, "If B is a knight then I am a knave."

What are A and B?

114. _____

Two individuals, X and Y, were being tried for participation in a robbery. A and B were court witnesses, and each of A,B is either a knight or a knave. The witnesses make the following statements:

> A: If X is guilty, so is Y.
> B: Either X is innocent or Y is guilty.

Are A and B necessarily of the same type? (We recall that two people from the island of knights and knaves are said to be of the same type if they are either both knights or both knaves.)

115. _____

On the island of knights and knaves, three inhabitants A,B,C are being interviewed. A and B make the following statements:

> A: B is a knight.
> B: If A is a knight, so is C.

Can it be determined what any of A,B,C are?

B. LOVE AND LOGIC _____

116. _____

Suppose the following two statements are true:

> (1) I love Betty or I love Jane.
> (2) If I love Betty then I love Jane.

Does it necessarily follow that I love Betty? Does it necessarily follow that I love Jane?

117. _____

Suppose someone asks me, "Is it really true that if you love Betty then you also love Jane?" I reply, "If it is true, then I love Betty."

Does it follow that I love Betty? Does it follow that I love Jane?

118. _____

This time we are given two girls, Eva and Margaret. Some-one asks me, "Is it really true that if you love Eva then you also love Margaret?" I reply, "If it is true, then I love Eva, and if I love Eva, then it is true."

Which girl do I necessarily love?

119. _____

This time we are given three girls, Sue, Marcia, and Dianne. Suppose the following facts are given:

(1) I love at least one of the three girls.
(2) If I love Sue but not Dianne, then I also love Marcia.
(3) I either love both Dianne and Marcia or I love neither one.
(4) If I love Dianne, then I also love Sue.

Which of the girls do I love?

Discussion. Aren't logicians a bit silly? Shouldn't I know whether or not I love Betty, Jane, Eva, Margaret, Sue, Marcia, Dianne, etc., without having to sit down and figure it out? Wouldn't it be funny if a wife asked her academic husband, "Do you love me?" and he answered, "Just a minute dear," and sat down for half an hour, calculating with paper and pencil, and then replied, "Yes, it turns out that I do"?

I am reminded of the allegedly true story of the philosopher Leibniz who was once wondering whether to marry a certain lady. He sat down with paper and pencil and made two lists, one list of advantages and one list of disadvantages. The second list turned out to be longer, so he decided not to marry her.

120.

This problem, though simple, is a bit surprising.

Suppose it is given that I am either a knight or a knave. I make the following two statements:

(1) I love Linda.
(2) If I love Linda then I love Kathy.

Am I a knight or a knave?

121. A Variant of an Old Proverb.

An old proverb says: "A watched kettle never boils." Now, I happen to know that this is false; I once watched a kettle over a hot stove, and sure enough it finally boiled. Now, what about the following proverb?

"A watched kettle never boils unless you watch it." Stated more precisely, "A watched kettle never boils unless it is watched."

Is this true or false?

C. IS THERE GOLD ON THIS ISLAND?

The puzzles of the last two groups were concerned largely with conditional statements—statements of the form "If P is true, so is Q." The puzzles of this group will be concerned largely with so-called biconditional statements— statements of the form "P is true if and only if Q is true." This statement means that if P is true then so is Q, and if Q is true then so is P. It means, in other words, that if either one of P, Q is true, so is the other. It also means that P, Q are either both true or both false. The statement "*P* if and only if *Q*" is symbolically written: "*P* ↔ *Q*."

The truth-table for *P* ↔ *Q* is this:

P	Q	P ↔ Q
T	T	T
T	F	F
F	T	F
F	F	T

The statement "*P* if and only if *Q*" is sometimes read "*P* is equivalent to *Q*" or "*P* and *Q* are equivalent." We note the following two facts:

F_1: Any proposition equivalent to a true proposition is true.

F_2: Any proposition equivalent to a false proposition is false.

122. Is There Gold on This Island? _____

On a certain island of knights and knaves, it is rumored that there is gold buried on the island. You arrive on the island and ask one of the natives, A, whether there is gold on this island. He makes the following response: "There is gold on this island if and only if I am a knight."

Our problem has two parts:

(a) Can it be determined whether A is a knight or a knave?
(b) Can it be determined whether there is gold on the island?

123. _____

Suppose, instead of A having volunteered this information, you had asked A, "Is the statement that you are a knight equivalent to the statement that there is gold on this island?" Had he answered "Yes," the problem would have reduced to the preceding one. Suppose he had answered "No." Could you then tell whether or not there is gold on the island?

124. How I·Became Rich. ————————————

This story is unfortunately not true. But it is an interesting story, so I will tell it to you anyway.

I found out about three neighboring islands A,B,C. I knew that there was gold buried on at least one of the three islands, but I didn't know which ones. Islands B and C were uninhabited; island A was inhabited by knights and knaves, and there was a possibility that there were some normals on the island, but I didn't know whether there were any normals or not.

I had the good fortune to find the map of the islands left by the famous, but capricious, Captain Marston—the pirate who had buried the gold. The message, of course, was in code. When decoded, it was seen to consist of two sentences. Here is the transcription:

> (1) THERE IS NO GOLD ON ISLAND A.
> (2) IF THERE ARE ANY NORMALS ON ISLAND A, THEN THERE IS GOLD ON TWO OF THE ISLANDS.

Well, I rushed over to island A; I knew the natives there knew all about the gold situation. The King of the island guessed what I was up to and told me in no uncertain terms that I would be allowed to ask only one question of any native I chose at random. I would have no way of knowing whether the native was a knight, knave, or normal.

My problem was to think of a question such that upon hearing the answer, I could then point to one of the islands and be sure there was gold on that island.

What question should I ask?

---

125. _____

Another time I was visiting a different island of knights, knaves, and normals. It was rumored that there was gold on the island, and I wanted to find out whether there was. The King of the island, who was a knight, graciously introduced me to three of the natives, A,B,C, and told me that at most one of them was normal. I was allowed to ask two yes-no questions to whichever ones I wished.

Is there a way of finding out in two questions whether there is gold on the island?

126. An Inferential Puzzle. _____

Suppose there are two neighboring islands each exclusively inhabited by knights and knaves (there are no normals). You are told that on one of the two islands there is an even number of knights and on the other one there is an odd number of knights. You are also told that there is gold on the island containing the even number of knights, but there is no gold on the other island.

You pick one of the two islands at random and visit it. All the inhabitants know how many knights and how many knaves live on the island. You are interviewing three inhabitants, A,B,C, and they make the following statements:

A: There is an even number of knaves on this island.
B: Right now, there is an odd number of people on the island.
C: I am a knight if and only if A and B are of the same type.

Assuming that you are neither a knight nor a knave and that at the moment you are the only visitor on the island, is there gold on the island or not?

SOLUTIONS

109–112. _____

These four problems all embody the same basic idea, which is that given any proposition P, if any person A on the island of knights and knaves says, "If I'm a knight then P," then the speaker must be a knight and P must be true! This is quite surprising, and we can prove this in two ways:

(1) Suppose that A is a knight. Then the statement "If A is a knight then P" must be a true statement (since knights always tell the truth). So A is a knight and it is true that if A is a knight then P. From these two facts it follows that P must be true. Thus the assumption that A is a knight leads to P as a conclusion. Therefore (recalling Fact 4 of Implication), we have proved that if A is a knight then P. But this is precisely what A asserted! Therefore A must be a knight. And since we have just proved that if A is a knight then P, then it follows that P must be true.

(2) An alternative way of seeing this is the following. We recall that a false proposition implies any proposition. Therefore if A is not a knight, then the statement, "If A is a knight then P" is automatically a true statement. Hence a knave would never make such a statement. So if a person who is either a knight or a knave makes such a statement, he can only be a knight and P must be true.

Let us apply this principle to our puzzles. As for 109, if we take P to be the proposition that B is a knight, then we see that A must be a knight and his statement is true, hence B is a knight. Thus the answer to 109 is that A and B are both knights.

As for 110, we take for P the proposition that A will eat his hat. We see that A must be a knight and that he must eat his hat. (This proves, incidentally, that knights, though doubtless virtuous and honorable, can sometimes be a bit stupid!)

As for 111, the answer again is that A is a knight.

As for 112, the correct conclusion is that the author is again spoofing! The problem is a paradox; no knight could make such a statement, nor could a knave either.

113.

A must be a knight and B must be a knave. To prove this, we first must prove that only a knight can make a statement of the form "If P, then I am a knave." As we recall, a true proposition is implied by any proposition; hence if the statement "I am a knave" is true, then so is the complete statement "If P, then I am a knave." But if I am a knave, I could never make that true statement. Hence if I say, "If P, then I am a knave," then I must be a knight.

Therefore A must be a knight. Hence also it is true that if B is a knight then A is a knave (because A says it is true). Then B can't be a knight, since this would imply that A is a knave, which he isn't.[1] Hence B is a knave.

114.

A, in effect, is saying that it is not the case that X is guilty and Y is innocent. This is but another way of saying that either X is innocent or Y is guilty, so A and B are really saying the same thing in different words. Therefore the two statements are either both true or both false, so A and B must be of the same type.

115.

Suppose A is a knight. Then so is B (since A says he is). Then B's statement—"If A is a knight, so is C"—is true. But A *is* a knight (by assumption), therefore C is a knight (under the assumption that A is).

[1] Any proposition which implies a false proposition must be false, since a true proposition can never imply a false proposition. In the above case, the proposition that B is a knight implies the false proposition that A is a knave, hence it must be false that B is a knight. This is another case of *reductio ad absurdum*.

We have just shown that if A is a knight, so is C.[2] Well, B said just that, hence B is a knight. Then A's statement that B is a knight is true, so A is also a knight. And we have shown that if A is a knight, so is C. Therefore C is a knight too. Therefore all three are knights.

116. _____

It doesn't follow that I love Betty but it does follow that I love Jane. To see that I love Jane, we reason as follows.

Either I love Betty or I don't. If I don't love Betty, then by condition (1), it must be Jane that I love (since it is given that I love at least one of them). On the other hand, if I love Betty, then by condition (2) I must love Jane as well. So in either case (whether I love Betty or whether I don't), it follows that I love Jane.

Incidentally, any female reader who happens to have the name "Betty" shouldn't be worried; just because it doesn't follow from the given conditions that I love Betty, it does not mean that it follows that I *don't* love Betty! It is quite possible that I love Betty too—maybe even more than Jane.

117. _____

This time it follows, not that I love Jane, but that I love Betty. For suppose I don't love Betty. Then the statement "If I love Betty then I love Jane" must be a true statement (since a false proposition implies any proposition). But it is given that if that statement is true, then I must love Betty. Therefore if I don't love Betty, it follows that I do love Betty, which is a contradiction. The only way out of the contradiction is that I do love Betty.

It cannot be determined whether or not I love Jane.

[2]We did this by assuming as a premise that A is a knight and drawing as a conclusion that C is a knight. By fact 4 of implication it follows that if A is a knight then C is a knight.

118.

It follows that I must love both girls. Let P be the statement, "If I love Eva then I also love Margaret. We are given:

(1) If P is true, then I love Eva.
(2) If I love Eva, then P is true.

We have seen in the solution of the preceding problem that from (1) it follows that I love Eva. Therefore I do love Eva. Therefore by (2), P must be true—i.e., it is true that if I love Eva I also love Margaret. But I do love Eva. Therefore I also love Margaret.

119.

I must love all three girls. There are several ways to prove this; here is one:

By (3) I either love both Dianne and Marcia or I love neither. Suppose I love neither. Then by (1) I must love Sue. Therefore I love Sue but not Dianne, and I don't love Marcia. This contradicts statement (2). Therefore it is not the case that I love neither Dianne nor Marcia, hence I love them both. Since I love Dianne, then by (4) I also love Sue. So I love all three.

120.

I must be a knight. If I were a knave, then both (1) and (2) would have to be false. Suppose (2) were false. Then I would love Linda but not Kathy, hence I would love Linda. This means that (1) would be true. So it is impossible that (1) and (2) are both false, hence I cannot be a knave.

121.

To say "P is false unless Q" is but another way of saying "If P then Q." (For example, to say, "I won't go to the movies unless you go with me" is equivalent to saying, "If I go to the movies, then you will go with me.") Thus the statement "A watched kettle never boils unless it is watched" is but another way of saying, "If a watched kettle boils, then it is watched." This, of course, is true, since a watched kettle is certainly watched, whether it boils or not.

122. _____

It is not possible to determine whether the speaker is a knight or a knave; nevertheless there must be gold on the island.

For purposes of this and the other problems of this section, let us establish once and for all the following basic principle: If a speaker (who is either a knight or a knave) makes the statement, "I am a knight if and only if P," then P must be true (regardless of whether the speaker is a knight or a knave).

To see this, let K be the proposition that the speaker is a knight. The speaker says that K is equivalent to P. Suppose the speaker is indeed a knight. Then K really is equivalent to P, and also K is true. Then P is equivalent to a true statement, hence P must be true. On the other hand, suppose the speaker is a knave. Then his statement is false, so P is not equivalent to K. Also, since he is a knave, K is false. Since P is not equivalent to the false proposition K, then P must be true (for if it were false, then it *would* be equivalent to K). Thus, whether the speaker is a knight or a knave, P must be true.

It is of interest to compare this with a principle established in the last section: If a knight or knave says, "If I'm a knight then P," then we can conclude that he is a knight and that P is true. But if a knight or knave says, "I am a knight if and only if P," then we can conclude that P is true, but we cannot determine whether or not he is a knight.

123. _____

Yes, you could; in this case there is no gold on the island.

Let G be the statement that there is gold on the island, and let K again be the statement that the speaker is a knight. The speaker, by answering "No," is asserting that G is *not* equivalent to K. Well, suppose the speaker is a knight. Then it really is the case that G is not equivalent to K. Now, since he is a knight, K is true. Therefore G, since it is not equivalent to the true proposition K, must be false. On the other hand, suppose he is a knave. Then G actually is equivalent to K (since the knave said they were not equivalent). But K is false (since the speaker is a knave). Thus G, being eqivalent to the false proposition K, must be false. So, whether the speaker is a knight or a knave, his "No" answer to the question indicates that G is false. So there is no gold on the island.

Discussion. The last two problems jointly imply a very important principle well known to "knight-knave" experts. As seen in the solutions of the last two problems, if P is any statement at all, whose truth or falsity you wish to ascertain, if a person known to be a knight or knave knows the answer to P, then you can find out from him in just one question whether P is true or false. You just ask him, "Is the statement that you are a knight equivalent to the statement that P is true?" If he answers "Yes," then you know that P is true; if he answers "No," then you know that P is false.

This principle will be used in the solution of the next three problems; we shall refer to it as the *fundamental principle.*

124. _____

We know in advance that there is no gold on island A, there is gold on island B or island C, and if anyone on island A is normal, then there is gold on both island B and island C.

Well, the question I asked the speaker was: "Is the statement that you are a knight equivalent to the statement that there is gold on island B?"

Suppose he answers "Yes." If he is either a knight or a knave, then there is gold on island B (by the fundamental principle established in the solution of the preceding problem). If he is normal, then again there is gold on islands B and C, so there is certainly gold on island B. Thus a "Yes" answers means that there is gold on island B.

Suppose he answers "No." If he is a knight or a knave, then there is no gold on island B (again by the fundamental principle). This means that there must be gold on island C. On the other hand, if he is normal, then there is gold on both island B and island C, so there is gold on island C. Thus a "No" answer means that there is gold on island C.

125.

This problem is solved by two uses of the fundamental principle (see solution of problem 123 for an explanation of the fundamental principle).

In one question it is possible to locate one of the three who you know is definitely *not* normal. You do this by asking A, "Is the statement that you are a knight equivalent to the statement that B is normal?" Suppose he answers "Yes." Is A is either a knight or a knave, then B must be normal (by the fundamental principle). This means that C is not normal. If A is not a knight or a knave, then he must be normal, so again C can't be normal. Thus a "Yes" answer means that C is not normal.

Suppose A answers "No." If he is a knight or a knave, then B is not normal (again by the fundamental principle). If A is not a knight or a knave, then again B is not normal, because A is. Thus a "No" answer means that B is not normal.

So, if you get a "Yes" answer from A, then you pick C to ask your second question; if you get a "No" answer, then you pick B. Thus you know you are questioning someone who is either a knight or a knave. You then ask him the same question as in problem 122, namely, whether the statement that he is a knight is equivalent to the statement that there is gold on the island. A "Yes" answer means that there is gold; a "No" answer means that there isn't.

126.

If you didn't know the fundamental principle, this problem would be most baffling. But now that you know the fundamental principle (see solution of problem 123), the problem is quite easy. I presume that you know that the sum of two even whole numbers is an even number, and the sum of two odd numbers is again even. This means that if you substract an even number from an even number you will get an even number, and if you subtract an odd number from an odd number, you will again get an even number. (For example, $12-8=4$; $13-7=6$.)

From C's statement it follows (by the fundamental principle) that A and B really are of the same type, i.e., they are both knights or both knaves. Thus their statements are either both true or both false. Suppose they are both true. Then by A's statement, there are an even number of knaves on the island. By B's statement there are an odd number of people including yourself. But you are neither a knight nor a knave, and the only visitor on the island, hence there are an even number of natives on the island. So, subtracting the even number of knaves from the even number of knights and knaves, you get an even number of knights. So in this case, there is gold on the island. On the other hand, suppose both statements are false. This means that there are an odd number of knaves on the island and an odd number of knights and knaves (an even number of people, including yourself). Then again there must be an even number of knights, so again there is gold on the island.

9. Bellini or Cellini?

This is a sequel to the story of Portia's caskets. We recall that whenever Bellini fashioned a casket he always wrote a true inscription on it, and whenever Cellini fashioned a casket he always wrote a false inscription on it. Now, Bellini and Cellini had sons who were also casket makers. The sons took after their fathers; any son of Bellini wrote only true statements on those caskets he fashioned, and any son of Cellini wrote only false statements on his caskets.

Let it be understood that the Bellini and Cellini families were the only casket makers of Renaissance Italy; all caskets were made either by Bellini, Cellini, a son of Bellini, or a son of Cellini.

It you should ever come across any of these caskets, they are quite valuable—especially those made by Bellini or Cellini.

A. WHOSE CASKET?

127. _____

I once came across a casket which bore the following inscription:

> THIS CASKET
> WAS NOT MADE BY
> ANY SON OF
> BELLINI

Who made this casket, Bellini, Cellini, a son of Bellini, or a son of Cellini?

128. _____

Another time I came across a casket whose inscription enabled me to deduce that the casket must have been made by Cellini.

Can you figure out what the inscription could have been?

129. _____

The most valuable caskets of all are those bearing an inscription such that one can deduce that the casket must have been made by Bellini or Cellini, but one cannot deduce which one. I once had the good fortune to come across such a casket. Can you figure out what the inscription could have been?

130. From the Sublime to the Ridiculous. _____

Suppose you came across a casket bearing the following inscription:

> THIS CASKET
> WAS MADE
> BY ME

What would you conclude?

131. A Florentine Nobleman. _____

A certain Florentine nobleman gave very lavish entertainments, the high point of which was a game in which the prize was a valuable jewel. This nobleman knew the story of Portia's caskets and designed his game accordingly. He had three caskets, gold, silver, and lead, and inside one of them was the jewel. He explained to the company that each of the caskets was made by Bellini or Cellini (and not any of their sons). The first person who could guess which casket contained the jewel, and who could prove his guess correct, would be awarded the jewel. Here are the inscriptions:

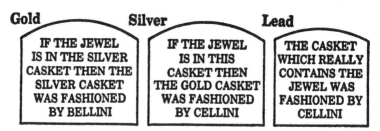

Gold

IF THE JEWEL
IS IN THE SILVER
CASKET THEN THE
SILVER CASKET
WAS FASHIONED
BY BELLINI

Silver

IF THE JEWEL
IS IN THIS
CASKET THEN
THE GOLD CASKET
WAS FASHIONED
BY CELLINI

Lead

THE CASKET
WHICH REALLY
CONTAINS THE
JEWEL WAS
FASHIONED BY
CELLINI

Which casket contains the jewel?

B. CASKET PAIRS

In some museums can be found pairs of caskets—one gold and one silver—made and originally sold as sets. Actually, the Bellini and Cellini families were the closest of friends and would sometimes collaborate in making a pair. Of course, only one person would make any one casket, but given a pair, it sometimes happened that one person made one of the caskets and another person made the other. The two families had great fun designing pairs such that intelligent posterity could figure out, or partly figure out, who were the makers. Given any set, there are sixteen possibilities: the gold casket could have been made by Bellini, a son of Bellini, Cellini, or a son of Cellini, and with each of these four possibilities there were four possibilities for the maker of the silver casket.

132. _____

I once came across the following pair:

Gold

BOTH CASKETS OF
THIS SET WERE MADE
BY MEMBERS OF THE
CELLINI FAMILY

Silver

NEITHER OF THESE
CASKETS WAS MADE
BY ANY SON OF BELLINI
OR ANY SON OF CELLINI

Who made each casket?

133.

I once came across the following pair:

Gold

IF THIS CASKET WAS
MADE BY ANY MEMBER
OF THE BELLINI FAMILY
THEN THE SILVER CASKET
WAS MADE BY CELLINI

Silver

THE GOLD
CASKET WAS MADE
BY A SON OF
BELLINI

Who made each casket?

134.

Consider the following pair:

Gold

THE SILVER CASKET
WAS MADE BY A
SON OF BELLINI

Silver

THE GOLD CASKET
WAS NOT MADE BY
A SON OF BELLINI

Prove that at least one of them was made by Bellini.

135.

Consider the following pair:

Gold

THE SILVER
CASKET WAS MADE
BY CELLINI

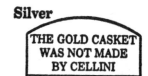

Silver

THE GOLD CASKET
WAS NOT MADE
BY CELLINI

Prove that at least one of the caskets was made by a son of Cellini.

136.

Consider the following pair:

Gold

THE SILVER CASKET
WAS MADE BY A
SON OF BELLINI

Silver

THE GOLD CASKET
WAS MADE BY A
SON OF CELLINI

Prove that at least one of the caskets was made by Bellini or Cellini.

137.

The next adventure I had was particularly remarkable. I came across a pair of caskets and I was interested to know whether at least one of them was fashioned by Bellini. I read the inscription on one of them, but I could not tell from it whether at least one of them was made by Bellini. Then I looked at the other inscription, which to my amazement was the same as the first, and to my further amazement, I could then tell that both caskets must have been made by Bellini.

Can you figure out what these inscriptions could have been?

138.

Another time I came across a pair bearing identical inscriptions from which I was able to infer that both caskets were made by Cellini, but from neither casket alone could I have inferred that even one of them was made by Cellini.

Can you supply such an inscription?

139.

Another time I came across a pair bearing identical inscriptions from which I was able to infer that either they were both made by Bellini or both made by Cellini, but I couldn't tell which. Also, from neither casket alone could I have inferred this.

Can you supply such an inscription?

140.

The most valuable pair of caskets which one can find is one satisfying the following conditions:

- (1) From the inscriptions one can deduce that one of them was made by Bellini and the other by Cellini, but one cannot know which casket was made by whom.
- (2) From neither casket alone can one deduce that the pair is a Bellini—Cellini pair.

I once had the good fortune to come across such a pair. (I understand that it is the only such pair ever made.) Can you supply such a pair of inscriptions?

141. A Delightful Adventure.

Once in my bachelor days I was in Florence. I read an ad in the papers: WANTED—A LOGICIAN. (Fortunately it was printed in English; I can't read Italian.) Well, I went to the museum which had placed the ad, and I was told that a logician was needed to help straighten out a baffling mystery. Four caskets had been found, two gold and two silver. It was known that they formed two sets, but somehow the sets had gotten mixed up, so it was not known which gold casket went with which silver casket. I was shown the four caskets and was soon able to straighten out the difficulty, for which I received an excellent consultant fee. Not only that, but I was also able to establish which casket was made by whom, for which I received an additional bonus (consisting, among other things, of an excellent case of Chianti), and I also received a kiss of gratitude from one of the most charming ladies in Florence.[1]

Here are the four caskets:

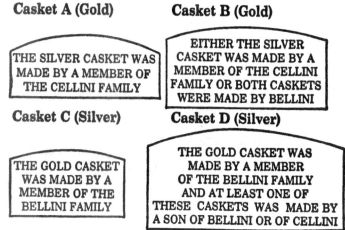

Casket A (Gold)

THE SILVER CASKET WAS MADE BY A MEMBER OF THE CELLINI FAMILY

Casket B (Gold)

EITHER THE SILVER CASKET WAS MADE BY A MEMBER OF THE CELLINI FAMILY OR BOTH CASKETS WERE MADE BY BELLINI

Casket C (Silver)

THE GOLD CASKET WAS MADE BY A MEMBER OF THE BELLINI FAMILY

Casket D (Silver)

THE GOLD CASKET WAS MADE BY A MEMBER OF THE BELLINI FAMILY AND AT LEAST ONE OF THESE CASKETS WAS MADE BY A SON OF BELLINI OR OF CELLINI

There are now two problems:

(a) Should A be paired with C or D?
(b) Who made each of the four caskets?

SOLUTIONS

127.

It was made by Bellini. If a son of Bellini had made the casket, the statement would be false, which is impossible. If

---

[1]Since Benvenuto Cellini was quite a braggart, why shouldn't I follow in his footsteps?

Cellini or a son of Cellini had made the casket, the statement would be true, which is impossible. Therefore it was made by Bellini.

128.

One inscription which would work is: This casket was made by a son of Cellini.

129.

"This casket was made by Bellini or a son of Cellini."

130.

The statement is obviously true, hence the casket was made by Bellini or a son of Bellini.

131.

Step One: Suppose the lead casket was made by Bellini. Then the statement on it is true, hence the jewel lies in a Cellini casket, so it cannot be in the lead casket. On the other hand, suppose the lead casket was made by Cellini. Then the statement on it is false, hence the jewel lies in a Bellini casket, hence is again not in the lead casket. This proves that the jewel does not lie in the lead casket.

Step Two: Next we know that the jewel cannot lie in the silver casket. If it did, we would get the following contradiction.

Suppose the jewel is in the silver casket. First, suppose the gold casket was made by Bellini. Then the statement on it is true, and since the jewel does lie in the silver casket (by assumption) then the silver casket is a Bellini. From this would follow that the gold casket was made by Cellini. So if the gold is a Bellini, then it is a Cellini.

On the other hand, suppose the gold casket is a Cellini. Then the statement on the gold casket is false, from which follows that the silver casket is not a Bellini, hence is a Cellini. Therefore the statement on the silver casket is false, from which follows that the gold casket is a Bellini. So if the gold casket is a Cellini, then it is a Bellini, which is impossible.

This proves that the jewel cannot be in the silver casket. Therefore it is in the gold casket.

132. _____

Clearly the statement on the gold casket cannot be true, or we would have a contradiction. So the gold casket was made by a member of the Cellini family. Since the statement is false, then not both caskets were made by members of the Cellini family, hence the silver casket was made by a member of the Bellini family. Therefore the statement on the silver casket is true, so neither casket was made by any of the sons. Therefore the gold casket was made by Cellini and the silver casket by Bellini.

133. _____

We recall that when an inhabitant of an island of knights and knaves says, "If I am a knight then so-and-so is true," then the inhabitant must be a knight and the so-and-so must be true. By a similar argument, we shall now show that the statement on the gold casket is true.

Suppose that the gold casket was made by a member of the Bellini family. Then the inscription on the gold casket is true: "If the gold casket was made by a member of the Bellini family, then the silver casket was made by Cellini." But the gold casket *was* made by a member of the Bellini family (this is our assumption), therefore the silver casket was made by Cellini. We have thus proved that if the gold casket was made by a member of the Bellini family then the silver casket was made by Cellini.[2] In other words, we have proved that the inscription on the gold casket is true. Therefore the gold casket was in fact made by a member of the Bellini family. This, together with the established fact that if the gold casket was made by a member of the Bellini family then the silver casket was made by Cellini, yields the fact that the silver casket was made by Cellini. Therefore the inscription on the silver casket is false, so the gold casket was not made by a son of Bellini. But the gold casket was made by a member of the Bellini family, therefore it was made by Bellini. So the gold casket was made by Bellini and the silver casket was made by Cellini.

134.

Suppose the statement on the gold casket is true. Then the silver casket was made by a son of Bellini, hence contains a true statement. This means that the gold casket was not made by a son of Bellini, but since the gold casket bears a true statement, then it must have been made by Bellini.

Suppose the statement on the gold casket is false. Then the silver casket was not made by a son of Bellini. However, the statement on the silver casket must be true (since the false statement on the gold casket could not have been made by a son of Bellini). So the silver casket was made by Bellini.

In summary, if the statement on the gold casket is true, then the gold casket was made by Bellini. If the statement on the gold casket is false, then the silver casket was made by Bellini.

135.

Suppose that the statement on the silver casket is true. Since it is a true statement, then the silver casket was made

[2]Because the premise that the gold casket was made by a member of the Bellini family led to the conclusion that the silver casket was made by Cellini. We have again made use of fact 4 of implication (see last paragraph of the preamble of Chapter 8).

by a member of the Bellini family, hence the statement on the gold casket—"the silver casket was made by Cellini"—must be false. But since the statement on the silver casket is true (by assumption), the gold casket was not made by Cellini. Therefore the gold casket contains a false statement but was not made by Cellini, hence it was made by a son of Cellini.

On the other hand, suppose the statement on the silver casket is false. Then the gold casket was made by Cellini, hence the statement on it is false, so the silver casket was not made by Cellini. Thus the silver casket contains a false statement but was not made by Cellini, so it was made by a son of Cellini.

136.

Suppose the gold casket inscription were true. Then the silver inscription would also have to be true, which would mean that the gold inscription was false. This is a contradiction, hence the gold inscription is false. This also means that the silver casket was not made by a son of Bellini. Therefore, if the silver inscription is true, then the silver casket was made by Bellini. If the silver inscription is false, then the gold casket was not made by a son of Cellini, but since the gold inscription is false, then the gold casket was made by Cellini.

In summary, if the silver inscription is true, then the silver casket was made by Bellini; if the silver inscription is false, then the gold casket was made by Cellini. So either the silver casket is a Bellini or the gold casket is a Cellini.

137.

There are many possible solutions to this and the next three problems. One solution for this problem is that both caskets contained the inscription: "Either both caskets were made by Bellini or at least one was made by a member of the Cellini family."

No member of the Cellini family could have made either of the caskets, because the statement would then be true. So both caskets were made by members of the Bellini family. The statements, therefore are true, so either both caskets were made by Bellini or at least one was made by a member of the Cellini family. The latter alternative is false, so both caskets are Bellini's.

138.

One solution is that both inscriptions read: "At least one of these caskets was made by a son of Cellini." If the statements were true, then at least one of the caskets would have been made by a son of Cellini, but it is not possible that a son of Cellini makes a true statement. Therefore the statements are false, which means that neither casket was made by a son of Cellini, hence Cellini made both caskets.

139.

An inscription which works is: "Either both caskets were made by Bellini or at least one was made by a son of Cellini."

We will prove that if the inscriptions are true, then both caskets were made by Bellini, and if the inscriptions are false, then both caskets were made by Cellini.

Suppose the inscriptions are true. Then it really is the case that either both caskets were made by Bellini or that at least one was made by a son of Cellini. The latter alternative is impossible (since a son of Cellini cannot write a true inscription), hence both caskets must have been made by Bellini.

Suppose the inscriptions are false. Then both alternatives of the disjunction are false—in particular the second alternative (that at least one was made by a son of Cellini) is false, which means that neither casket was made by a son of Cellini. Yet both inscriptions are false, so they were made by Cellini.

140. _____

One solution is the following:

> *Gold*: "These caskets were made by Bellini and Cellini if and only if the silver casket was made by a member of the Cellini family."
>
> *Silver*: "The gold casket was made by a member of the Cellini family."

We let P be the proposition that the caskets were made by Bellini and Cellini, and Q may be the proposition that the silver casket was made by a member of the Cellini family. The inscription on the gold casket says that P is equivalent to Q; the inscription on the silver casket says that the inscription on the gold casket was made by a liar, which in effect says that the inscription on the gold casket is false. This means that one of the two inscriptions is true and the other one is false.

Suppose the inscription on the gold casket is true. Then (since we have shown that one inscription is true and one is false), the inscription on the silver casket must be false, hence it was made by a member of the Cellini family, so Q is true. Also, since the inscription on the gold casket is true, then P really is equivalent to Q. Then (since Q is true) P must be true.

Suppose the inscription on the gold casket is false. Then the inscription on the silver casket is true, hence it was not made by any Cellini, so Q must be false, and also P is not equivalent to Q. Hence again P is true.

We see that in either case, P must be true, that is, one of the caskets was made by Bellini and the other by Cellini.

141. _____

Casket A must be paired with casket D, for if it were paired with casket C we would get the following contradiction.

Suppose A were paired with C. Suppose the inscription on A is true. Then the inscription on C is false. This means that the inscription on A is false. This is a contradiction. On the other hand, suppose the inscription on A is false. Then the inscription on C is true. This means that the inscription on A is true—again a contradiction. Therefore A is not paired with C. This solves the first half of the problem.

Now let us consider the B–C pair. Suppose the statement on C is false. Then B was made by a member of the Cellini family, hence contains a false statement. This means that neither alternative of the statement is true, hence the first alternative is false, which means that C was made by a member of the Bellini family. So, if the statement on C is false then C was made by a member of the Bellini family, which is impossible. Hence the statement on C is true. Therefore the statement on B is also true (because it says on C that B was made by a member of the Bellini family). Now, the first alternative of the statement on B cannot be true, therefore the second is. Thus caskets B and C were both made by Bellini.

Now let us consider the A–D pair. Suppose the inscription on A is false. Then D was made by a member of the Bellini family and hence the inscription on it is true. This would mean that A was made by a member of the Bellini family, so we would get a contradiction. Therefore the inscription on A is true. This further implies that the inscription on D is false. Hence at least one of the alternatives of the statement is false. The first alternative is true (since the statement on A is true), hence the second alternative is false. This means that neither casket was made by a son of Bellini or Cellini. Therefore A was made by Bellini and D was made by Cellini.

PART THREE
Weird Tales

10. The Island of Baal

A. IN QUEST OF THE ABSOLUTE

I read, in some philosophical textbook or other, "The true philosopher is the little girl of nine who was looking out of a window and suddenly turned to her mother and said, 'But mother, what puzzles me is how come there is anything at all?' "

This problem has baffled many a philosopher; some philosophers have regarded this as *the* fundamental philosophical problem. They put it in the form: "Why is there something instead of nothing?"

When you stop to think of it, it is really a good question, isn't it? Actually, why *is* there something instead of nothing? Well, once upon a time there was a certain philosopher who decided to make it the main project of his life to find out why there is something instead of nothing. First he read all the books on philosophy, but none of them could tell him the real reason why there is something instead of nothing. Next, he turned to theology. He asked all the learned rabbis, priests, bishops, ministers, and other religious leaders, but none of them could satisfactorily explain why there is something instead of nothing. Then he turned to Eastern philosophy; he went wandering around for twelve years in India and Tibet, interviewing various gurus, but none of them knew why there is something instead of nothing. Then he spent another twelve years in China and Japan interviewing various Taoist hermits and Zen-masters Finally he met one sage who was on his deathbed and who said:

"No, my son, I myself do not know why there is something instead of nothing. The only place on this planet where the answer is known is on the island of Baal. One of the high priests of the Temple of Baal knows the true answer."

"And where is the island of Baal?" asked the philosopher eagerly.

"Ah!" was the reply, "I don't know that either. In fact, I have never known anyone who has actually found his way to Baal. All I know about is the location of a certain uncharted cluster of islands on one of which is a map and a complete set of directions to the island of Baal. I do not know on which island of the cluster the map can be found; all I know is that it is on one of them, and the name of that one is "Maya." However, all these islands are inhabited exclusively by knights, who always tell the truth, and knaves, who always lie. Hence one has to be very cagey!"

This was the most promising news the philosopher had heard in twenty-four years! Well, he found his way without difficulty to this cluster and systematically tried one island after another, hoping to find out which one was the island of Maya.

142. The First Island. _____

On the first island he tried, he met two natives A,B, who made the following statements:

    A: B is a knight and this is the island of Maya.
    B: A is a knave and this is the island of Maya.

Is this the island of Maya?

143. The Second Island. _____

On this island, two natives A,B, made the following statements:

> A: We are both knaves, and this is the island of Maya.
> B: That is true.

Is this the island of Maya?

144. The Third Island. _____

On this island, A and B said:

> A: At least one of us is a knave, and this is the island of Maya.
> B: That is true.

Is this the island of Maya?

145. The Fourth Island. _____

On this island, two natives A,B said:

> A: Both of us are knaves, and this is the island of Maya.
> B: At least one of us is a knave, and this is not the island of Maya.

Is this the island of Maya?

146. The Fifth Island. _____

Two of the natives here, A and B, said:

> A: Both of us are knaves, and this is the island of Maya.

B: At least one of us is a knight, and this is not the island of Maya.

Is this the island of Maya?

147. The Sixth Island. _____

On this island, two of the natives A,B made the following statements:

A: Either B is a knight, or this is the island of Maya.
B: Either A is a knave, or this is the island of Maya.

Is this the island of Maya?

148. The Map to Baal. _____

Well, our philosopher found the island of Maya. However, the task of finding the map and directions to Baal was not as easy as he had anticipated. He had to see the High Priest of Maya. The priest led him into a room in which three maps X,Y,Z were lying on a table. The priest explained that only one of the maps was the true map to Baal; the other two maps each led to an island of demons, and if one landed on an island of demons, he would be instantly demolished. The philosopher had to choose one of the three maps.

Well, in the room were five witch doctors, A, B, C, D, and E. Each witch doctor was either a knight or a knave. They gave him the following bits of advice:

A: X is the correct map.
B: Y is the correct map.
C: A and B are not both knaves.
D: Either A is a knave or B is a knight.
E: Either I am a knave or C and D are of the same type (both knights or both knaves).

Which of the maps X,Y,Z is the correct one?

B. THE ISLAND OF BAAL

Of all the islands of knights and knaves, the island of Baal is the weirdest and most remarkable. This island is inhabited exclusively by humans and monkeys. The monkeys are as tall as the humans, and speak as fluently. Every monkey, as well as every human, is either a knight or a knave.

In the dead center of this island stands the Temple of Baal, one of the most remarkable temples in the entire universe. The high priests are metaphysicians, and in the Inner Sanctum of the temple can be found a priest who is rumored to know the answer to the ultimate mystery of the universe: why there is something instead of nothing.

Aspirants to the Sacred Knowledge are allowed to visit the Inner Sanctum, provided that they prove themselves worthy by passing three series of tests. I learned all these secrets, incidentally, by stealth: I had to enter the temple disguised as a monkey. I did this at great personal risk. Had I been caught, the penalty would have been unimaginable. Instead of merely annihilating me, the priests would have changed the very laws of the universe in such a way that I could never have been born!

Well, our philosopher chose the right map and arrived safely on the island of Baal and agreed to try the tests. The first series took place on three consecutive days in a huge room called the Outer Sanctum. In the center of the room a cowled figure was seated on a golden throne. He was either a human or a monkey, and also a knight or a knave. He uttered a sacred sentence, and from this sentence the philosopher had to deduce exactly what he was—whether a knight or a knave, and whether a human or a monkey.

149. The First Test. _____

The speaker said, "I am either a knave or a monkey."
Exactly what is he?

150. The Second Test. _____

The speaker said, "I am a knave and a monkey."
Exactly what is he?

151. The Third Test. _____

The speaker said, "I am not both a monkey and a knight."
What is he?

The philosopher passed these three tests, so he was allowed to try the second series, which also took place on three consecutive days and in another great room, known as the Middle Sanctum. In this room there were two cowled figures seated on platinum thrones. They uttered sacred sentences, and the philosopher was then to give a complete description of each speaker. We will call the speakers A and B.

152. The Fourth Test. _____

> *A*: At least one of us is a monkey.
> *B*: At least one of us is a knave.

What are A and B?

153. The Fifth Test. _____

> *A*: Both of us are monkeys.
> *B*: Both of us are knaves.

What are A and B?

154. The Sixth Test. _____

> *A*: B is a knave and a monkey. I am human.
> *B*: A is a knight.

What are A and B?

The philosopher passed the second series of tests and took the third series, which consisted of only one test, but it was a complicated one.

155.

There are four doors X,Y,Z,W leading out of the Middle Sanctum. At least one of them leads to the Inner Sanctum. If you enter a wrong door, you will be devoured by a fierce dragon.

    Well, there were eight priests A,B,C,D,E,F,G,H, each of whom is either a knight or a knave. They made the following statements to the philosopher:

    A: X is a good door.
    B: At least one of the doors Y,Z is good.
    C: A and B are both knights.
    D: X and Y are both good doors.
    E: X and Z are both good doors.
    F: Either D or E is a knight.
    G: If C is a knight, so is F.
    H: If G and I are both knights, so is A.

Which door should the philosopher choose?

156. In the Inner Sanctum!

The philosopher chose the correct door and safely entered the Inner Sanctum. Seated on two diamond thrones were the two greatest priests in the entire universe! It is possible that at least one of them knew the answer to the Great Question: "Why is there something instead of nothing?"

    Of course, each of the two great priests was either a knight or a knave. (Whether they were human or monkey is not relevant.) So we do not know of either whether he is a knight or a knave, or whether he knows the answer to the

Great Question. The two priests made the following statements:

> *First Priest* / I am a knave, and I don't know why there is something instead of nothing.
> *Second Priest* / I am a knight, and I don't know why there is something instead of nothing.

Did either of the priests really know why there is something instead of nothing?

157. The Answer!

And now you are about to find out the true answer to the Great Question of why there is something instead of nothing!

Well, one of the two priests, who in fact did know the answer to the Great Question, when asked by the philosopher, "Why is there something instead of nothing?" gave the following response:

"There is something instead of nothing."

What drastic conclusion follows from all this?

SOLUTIONS

142.

Suppose B is a knight. Then this is the island of Maya and also A is a knave. Hence A's statement is false, so it is not true that B is a knight and this is the island of Maya. However B is a knight by assumption. Hence the first part of the statement is true; therefore the second part of the statement is false, hence this is not the island of Maya. So if B is a knight, it follows that this island both is and is not the island of Maya. Therefore B must be a knave.

Since B is a knave, it follows that A is also a knave (because A claims that B is a knight). Since B is a knave, his statement is false, therefore it is not true that A is a knave and this is the island of Maya. But the first part of the statement is true (since A is a knave), therefore the second part must be false, hence this is not the island of Maya.

143.

Obviously A is a knave (a knight could never make A's statement). Since B agrees with A, then B is also a knave. Since A's statement is false, then it is not true that (1) they are both knaves and (2) this is the island of Maya. However, (1) *is* true, so (2) must be false. Therefore this island is not the island of Maya.

144.

Since B agrees with A, then they are either both knights or both knaves. If they were both knights then it would not be the case that at least one of them is a knave, hence A's statement would be false, which is impossible since A would be a knight. Therefore they are both knaves. This means that A's statement is false. But the first clause of A's statement must be true (they are both knaves so at least one of them is a knave), hence the second clause must be false. Therefore this is not the island of Maya.

145.

A is certainly a knave, since a knight couldn't make that statement. If B is a knight, then, by his statement, this is not the island of Maya. If B is a knave, then the first clause of A's statement is true; but A's statement is false, since A is a knave, hence the second clause must be false. So again, this is not the island of Maya.

146.

Again, A must be a knave, B can be either a knight or a knave, but in either case, this is not the island of Maya.

147.

If A were a knave, then both clauses of his disjunctive statement would be false, which would mean that B was a knave. This would mean that both clauses of B's disjunctive statement would be false, so A would be a knight. This is a contradiction; therefore A is a knight. Therefore his statement is true, hence either B is a knight or this is the island of Maya. If the second alternative is true, then, of course, this is the island of Maya. Suppose the first alternative is true, that is, suppose B is a knight. Then B's statement is true: "A is a knave, or this is the island of Maya." But A is not a knave, so the first alternative is false. Therefore the second alternative is true, so this is the island of Maya.

To repeat part of this argument, we have seen that either B is a knight or this is the island of Maya. But also, if B is a knight, then again this is the island of Maya. Therefore this is the island of Maya.

So we have found the island of Maya—at last!

148.

If E were a knave, then it would be true that either E is a knave or C and D are of the same type. This would mean that a knave made a true statement, which is impossible. Therefore E is a knight. Hence his statement is true, so either he is a knave or C and D are of the same type. But he is not a knave, hence C and D are of the same type.

Suppose C were a knave. Then A and B would both be knaves. Then D's statement would be true, hence D would be a knight. Thus C would be a knave and D a knight, which is contrary to the fact that C and D are of the same type. Therefore C must be a knight; hence D is also a knight. Since C is a knight, then A and B are not both knaves, hence either X or Y is the correct map. Suppose X were the correct map. Then A is a knight and B is a knave, contrary to D's true statement that either A is a knave or B is a knight. So X cannot be the correct map, so the correct map must be Y.

149.

If the speaker were a knave, then he would be either a knave or a monkey, hence his statement would be true, contrary to the fact that he is a knave. Therefore he is a knight. This means his statement is true, hence he is either a knave or a monkey. He is not a knave, therefore he is a monkey. Thus he is a monkey knight.

150.

Clearly the speaker is not a knight, hence he is a knave and his statement is false. Therefore he is either a knight or a human. He is not a knight, therefore he is human. Hence he is a human knave.

151.

Suppose the speaker were a knave. Then it would be true that he is not both a monkey and a knight, hence his statement would be true, and we would have a knave making a true statement. Therefore the speaker is a knight. Therefore it is true that he is not both a monkey and a knight. If he were a monkey, then he would be both a monkey and a knight. Hence he is human. So he is a human knight.

152.

B can't be a knave, or his statement would be true. Therefore B is a knight. Hence his statement is true, so A must be a knave. Then A's statement is false, so they are both human. Therefore A is a human knave and B is a human knight.

153.

B must be a knave, because a knight could not make that statement. Therefore not both A and B are knaves, so A is a knight. Hence A's statement is true, so both of them are monkeys. Hence A is a monkey knight and B is a monkey knave.

154. _____

Suppose B were a knight. Then A would be a knight (since B says he is), hence B would have to be a knave and a monkey, which is a contradiction. Therefore B is a knave. Hence, by B's statement, A is also a knave. Since A's first statement is false, B is not a knave and a monkey. But B *is* a knave, so it must be false that B is a monkey. So B is a human knave. From A's second statement it follows that A is a monkey. So A is a monkey knave.

155. _____

We will first show that G is a knight. To do this, it suffices to show that his statement is true. So we must show that if C is a knight, so is F. We do this by assuming that C is a knight, and then showing that F is also a knight.

Well, suppose C is a knight. Then A and B are both knights. Hence X is a good door and either Y or Z is good.

Case One: Y is good. Then X,Y are both good. In this case, D is a knight.

Case Two: Z is good. Then X,Z are both good. In this case, E is a knight.

Hence either D or E must be a knight. Therefore F's statement is true, so F is a knight.

Our assumption that C is a knight leads to the conclusion that F is a knight. Therefore it is true that if C is a knight, so is F. This is what G said, therefore G is a knight.

Now we will prove that H's statement is true. H said that if G and H are both knights, so is A. Suppose that H is a knight. Then G and H are both knights. Also it is true that if

G and H are both knights, so is A (because H said it was, and we are assuming that H is a knight). Therefore if H is a knight, then (1) G and H are both knights; (2) if G and H are both knights, so is A. From (1) and (2) it follows that A is a knight. So if H is a knight, so is A. This is what H said, so H must be a knight. Hence his statement is true, and since G and H are both knights, A is a knight.

Now we know that A is a knight. Hence X really is a good door. So the philosopher should choose door X.

156.

The first priest couldn't be a knight; he must be a knave. Hence his statement is false, which means that it is not true that he is a knave and that he doesn't know the answer to the Great Question. But he is a knave, so the first part of the statement is true. Therefore, the second part of the statement must be false, so he does know the answer. Therefore the first priest is both a knave and knows the answer.

As for the second priest, he is indeterminate; he is either a knight who doesn't know the answer or a knave. At any rate (and this is crucial for the next problem!) if he does know the answer, then he is a knave.

157.

We have seen that the first priest knows the answer to the question and is a knave, and the second priest, if he knows the answer, is a knave. We are given that the one who said "There is something instead of nothing" knew the answer. Therefore the one who said that is a knave, hence the statement "There is something instead of nothing" must be false! This means that nothing exists!

So, it appears that the answer to the philosopher's lifelong quest is that nothing really exists after all. However, there is one thing wrong; if nothing exists, then how come there was the priest who made the statement?

What properly follows, therefore, is that the island of

Baal, as I have described it, cannot exist. It's not merely the case that it *doesn't* exist (which was highly probable from the beginning of the story), but that it is logically certain that it *cannot* exist. For if it existed, and my story were true, then (as I have shown) it would logically follow that nothing exists, and hence the island of Baal wouldn't exist. This is a contradiction, hence the island of Baal cannot exist.

The curious thing is that up until the last story (problem 157), everything I told you, no matter how implausible it may have seemed, was logically possible. But when I told you the last story, that was the straw that broke the camel's back!

11. The Island of Zombies

A. "BAL" AND "DA"

On a certain island near Haiti, half the inhabitants have been bewitched by voodoo magic and turned into zombies. The zombies of this island do not behave according to the conventional concept: they are not silent or deathlike—they move about and talk in as lively a fashion as do the humans. It's just that the zombies of this island always lie and the humans of this island always tell the truth.

So far, this sounds like another knight-knave situation in a different dress, doesn't it? But it isn't! The situation is enormously complicated by the fact that although all the natives understand English perfectly, an ancient taboo of the island forbids them ever to use non-native words in their speech. Hence whenever you ask them a yes-no question, they reply "Bal" or "Da"—one of which means *yes* and the other *no*. The trouble is that we do not know which of "Bal" or "Da" means *yes* and which means *no*.

158. _____

I once met a native of this island and asked him, "Does 'Bal' mean *yes*?" He replied, "Bal."

(a) Is it possible to infer what "Bal" means?

(b) Is it possible to infer whether he is a human or a zombie?

159. _____

If you meet a native on this island, is it possible in only one question to find out what "Bal" means? (Remember, he will answer "Bal" or "Da".)

160. _____

Suppose you are not interested in what "Bal" means, but only in whether the speaker is a zombie. How can you find this out in only one question? (Again, he will answer "Bal" or "Da.")

161. Making the Medicine Man say "Bal." _____

You are on this same island and wish to marry the King's daughter. The King wants his daughter to marry only someone who is very intelligent, hence you have to pass a test.

The test is that you may ask the medicine man any one question you like. If he answers "Bal" then you may marry the king's daughter; if he answers "Da" then you may not.

The problem is to design a question such that regardless of whether the medicine man is a human or a zombie, and regardless of whether "Bal" means *yes* or *no*, he will have to answer "Bal."

162. _____

Here is a more difficult one. There is a rumor that there is gold on this island. You arrive on the island, and before you start excavating, you want to know whether there really is gold or not. The natives all know whether or not there is.

How, in one question to any one of the natives, can you find out? Remember, he will answer "Bal" or "Da," and from his answer you must know whether there is gold, regardless of what "Bal" and "Da" really mean.

B. ENTER INSPECTOR CRAIG

163. A Trial.

On a neighboring island of humans and zombies "Bal" and "Da" are again the native words for *yes* and *no*, but not necessarily in that order. Some of the natives answer questions with "Bal" and "Da," but others have broken away from the taboo and answer with the English words "Yes" and "No."

For some odd reason, given any family on this island, all members are of the same type. In particular, given any pair of brothers, they are either both human or both zombies.

A native was suspected of high treason. The case was so important, that Inspector Craig had to be called over from London. The three key witnesses were A,B, and C—all natives of the island. The following transcript is from the court records; Inspector Craig did the questioning.

Question (to A) / Is the defendant innocent?
A's Answer / Bal.
Question (to B) / What does "Bal" mean?
B's Answer / "Bal" means *yes.*
Question (to C) / Are A and B brothers?
C's Answer / No.
Second Question to C / Is the defendant innocent?
C's Answer / Yes.

Is the defendant innocent or guilty?

164. _____

In the above problem, can it be determined whether A and B are of the same type?

165. Semi-zombies. _____

After the trial, Inspector Craig paid a visit to a curious neighboring island: Some of the natives were human, some were zombies, and the others were what is known as *semizombies*. These semi-zombies have been subjected to voodoo magic, but the magic spells were only partially successful. As a result, the semi-zombies sometimes lie and sometimes tell the truth. Again the native words for *yes* and *no* are "Bal" and "Da" (though not necessarily respectively). The natives sometimes answer yes-no questions in English and sometimes with "Bal" and "Da."

Inspector Craig met one of the natives and asked him the following question: "When someone asks you whether 'Bal' means *yes*, and you answer in your native tongue, do you answer 'Bal'?"

The native answered, but Inspector Craig failed to record the answer, nor did he record whether it was given in English or in the native tongue. All Inspector Craig did record was that from the answer, he was able to deduce whether the speaker was a human, a zombie, or a semi-zombie.

What answer did the speaker give, and was it in English or in his native tongue?

166. Which? _____

Another time on the same island, Inspector Craig asked another native the following question: "When someone asks you whether two plus two equals four, and you answer in your native tongue, do you answer 'Bal'?"

Again, Inspector Craig did not record whether the answer was "Bal," "Da," "Yes," or "No," but again he could deduce whether the speaker was a human, a zombie, or a semi-zombie.

What answer did he get?

SOLUTIONS

158.

It is not possible to tell what "Bal" means, but we can tell that the speaker must have been human.

Suppose "Bal" means *yes*. Then "Bal" is the truthful answer to the question whether "Bal" means *yes*. So in this case, the speaker was human.

Suppose "Bal" means *no*. Then "No" is the truthful English answer to the question whether "Bal" means *yes*, therefore "Bal" is the truthful native answer to the question. So again, the speaker is human. So, regardless of whether "Bal" means *yes* or *no*, the speaker is human.

159.

All you have to ask him is whether he is human. All natives of this island claim to be human, so both a human and a zombie will answer affirmatively. So if he answers "Bal," then "Bal" means *yes*; if he answers "Da," then "Da" means *yes* (and "Bal" means *no*).

160.

The question of problem 158 does the job; just ask him if "Bal" means *yes*. If "Bal" does mean *yes*, then "Bal" is the correct answer to the question, so a human will say "Bal" and a zombie will say "Da." If "Bal" does not mean *yes*, then again "Bal" is the correct answer to the question, so again a human will say "Bal" and a zombie will say "Da."

161.

There are several ways to do this. One way is to ask the medicine man whether "Bal" is the true answer to the question of whether he is human. We can prove that he must answer "Bal." To simplify the exposition a bit, let H be the question, "Are you human?" Remember, you are not asking him whether H is true or false, but whether "Bal" is the correct answer to H.

Case One: He is human. If "Bal" means *yes*, then "Bal" is the correct answer to H, and since he is human, he will truthfully tell you that it is, hence he will say "Bal." If "Bal" means *no*, then "Bal" is not the correct answer to H, hence he will truthfully tell you it isn't, so he will say "Bal" (meaning *no*). Thus a human will answer "Bal" regardless of whether "Bal" means *yes* or *no*.

Case Two: He is a zombie. If "Bal" means *yes* then "Bal" is not the correct answer to H, but since he is a zombie, he will lie and say that it is the correct answer, so he will say "Bal" (meaning "Yes, it is the correct answer," which of course is a lie). If "Bal" means *no*, then "Bal" is the correct answer to H, hence he will lie and say it is not the correct answer, so he will say "Bal" (meaning *no*). So a zombie will say "Bal" regardless of whether "Bal" means *yes* or *no*.

There are other questions which would also do the job. Here are some:

(1) Is it the case that either you are human and "Bal" means *yes*, or that you are a zombie and "Bal" means *no*?

(2) Is it the case that you are human if and only if "Bal" means *yes*?

162.

Again, there are several ways to do this. One way is to ask, "If someone asked you whether there is gold on this island, would you answer 'Bal'?" As we will show, if there is gold on the island, then he will answer "Bal," and if there isn't, then he will answer "Da," regardless of whether he is human or a zombie and regardless of what "Bal" and "Da" really mean.

We let G be the question, "Is there gold on this island?

Case One: He is human and "Bal" means yes. Suppose there is gold on the island. Then he would answer "Bal" to the question G. Being human, he would truthfully tell you that he would answer "Bal," so he answers "Bal" to your question. Suppose there is no gold on the island. Then he wouldn't answer "Bal" to question G, and being human he would tell you that he wouldn't, so he answered "Da" to your question.

Case Two: He is a zombie and "Bal" means yes. Suppose there is gold on the island. Then again "Bal" is the truthful answer to G, so he, being a zombie, wouldn't answer "Bal" to G. But then he would lie to you and tell you that he would answer "Bal" to G. So his answer to you is "Bal." Suppose there is no gold on the island. Then "Bal" is a false answer to G, so he would in fact give that answer to G. But then he would lie to you and say that he wouldn't say "Bal," so he answers your question with "Da."

Case Three: He is human and "Bal" means no. Suppose there is gold on the island Then "Bal" is the false answer to G, so a human wouldn't make it. Then he would truthfully tell you that he wouldn't say "Bal," so he answers your question with "Bal." If there isn't gold on the island, then "Bal" is the truthful answer to G, hence is the answer the human would actually give to G. So he answers your question with "Da" (meaning "Yes, I would answer 'Bal' to G").

Case Four: He is a zombie and "Bal" means no. Suppose there is gold on the island. Then he would actually answer

"Bal" to G, but he would tell you that he wouldn't, so he will answer your question with "Bal." Suppose there isn't gold on the island. Then he would actually answer "Da" to G; he wouldn't answer "Bal" to G. But he would tell you that he would. Hence he answers "Da" to your question.

In summary, if there is gold on the island, then in each of the four cases you will get "Bal" for an answer; if there is no gold, you will get "Da" for an answer.

Another question that would work is this: "Is it the case that you are human if and only if 'Bal' is the true answer to whether there is gold on this island?"

163. _____

I shall first prove that C cannot be a zombie. Well, suppose he were. Then A and B must be brothers, hence both human or both zombies. Suppose they are both human. Then "Bal" really does mean *yes*, hence A in effect answered *yes* to whether the defendant is innocent, so the defendant is innocent. Suppose A,B are both zombies. Then "Bal" really means *no*, and since A is a zombie and answered *no* to whether the defendant is innocent, then the defendant is innocent. So if C is a zombie then the defendant is innocent (regardless of whether A,B are both human or both zombies). On the other hand, if C is a zombie then the defendant must be guilty, since C says he is innocent. This is a contradiction; therefore C can't be a zombie, so he is human. And since C says the defendant is innocent, then the defendant really is innocent.

164. _____

Since C is human, then A,B are not brothers. This, of course, does not necessarily mean that they are of different types; they may be of the same type even though they are not brothers. As a matter of fact, they *must* be of the same type, for if they were of different types, then the defendant

---

would have to be guilty. The reader should easily be able to prove this himself.

165.

Of all the four possible answers—"Bal," "Da," "Yes," "No"—the only one which neither a human nor a zombie could make is "No." More specifically, if the speaker were either one, had he answered in English, his answer would have to be "Yes"; if he answered in his native tongue, then if "Bal" means *yes*, he would have answered "Bal" (regardless of whether he is a human or a zombie), and if "Bal" means *no*, he would have answered "Da." (I leave it to the reader to prove these facts.) Therefore if Craig had gotten any answer but "No," he couldn't have known what the speaker was. But he did know, hence he got the answer "No," and the speaker was a semi-zombie.

166.

Again, the speaker must be a semi-zombie, and the only way Craig could know what the speaker was is by getting the answer "Da." Had the speaker answered in English, Craig could not have known, for both a human and a zombie would have answered "Yes" if "Bal" means yes, and "no" if "Bal" means *no*. Had the speaker answered "Bal," he could have been either a human, a zombie or a semi-zombie.

12. Is Dracula Still Alive?

A. IN TRANSYLVANIA

Despite what Bram Stoker has told us, I had grave reason to doubt that Count Dracula was ever really destroyed. I accordingly decided to go to Transylvania to investigate the truth for myself. My purposes were: (1) to ascertain whether Count Dracula was still alive; (2) in the event that he was destroyed, I wished to see his actual remains; (3) in the event that he was still alive, I wished to meet him.

At the time I was in Transylvania, about half the inhabitants were human and half were vampires. The humans and vampires are indistinguishable in their outward appearance, but the humans (at least in Transylvania) always tell the truth and the vampires always lie. What enormously complicates the situation is that half the inhabitants of Transylvania are totally insane and completely deluded in their beliefs—all true propositions they believe to be false and all false propositions they believe to be true. The other half are completely sane and know which propositions are true and which ones false. Thus the inhabitants of Transylvania are of four types: (1) sane humans; (2) insane humans; (3) sane vampires; (4) insane vampires. Whatever a sane human says is true; whatever an insane human says is false; whatever a sane vampire says is false;

and whatever an insane vampire says is true. For example, a sane human will say that two plus two equals four; an insane human will say it doesn't (because he really believes it doesn't); a sane vampire will also say it doesn't (because he knows it does and then lies); an insane vampire will say it does (because he believes it doesn't, and then lies about what he believes).

167.

I once met a Transylvanian who said, "I am human or I am sane."

Exactly what type was he?

168.

Another inhabitant said, "I am not a sane human."
What type was he?

169.

Another inhabitant said, "I am an insane human."
Is he of the same type as the last inhabitant?

170.

I once met an inhabitant and asked him, "Are you an insane vampire?" He answered "Yes" or "No," and I knew what he was.

What was he?

171.

I once met a Transylvanian who said, "I am a vampire."
Can it be inferred whether he is human or a vampire?
Can it be inferred whether he is sane?

172. _____

Suppose a Transylvanian says, "I am insane."

    (a) Can it be inferred whether he is sane?
    (b) Can it be inferred whether he is a human or a vampire?

173. An Ingenious Puzzle. _____

The converse of a statement "If P then Q" is the statement "If Q then P." Now, there exist two statements X, Y which are converses of each other and such that:

    (1) Neither statement is deducible from the other.
    (2) If a Transylvanian makes either one of the statements, it follows that the other one must be true.

Can you supply two such statements?

174. _____

Given any statement X, suppose a Transylvanian believes that he believes X. Does it follow that X must be true? Suppose he doesn't believe that he believes X. Does it follow that X must be false?

175. _____

Suppose a Transylvanian says, "I believe X." If he is human, does it follow that X must be true? If he is a vampire, does it follow that X must be false?

    The answer to this problem constitutes an important general principle!

176. _____

I once met two Transylvanians, A and B. I asked A, "Is B human?" A replied, "I believe so." Then I asked B, "Do you believe A is human?" What answer did B give (assuming he answered "Yes" or "No")?

177. _____

Let us define a Transylvanian to be *reliable* if he is either a sane human or an insane vampire and to be *unreliable* if he is either an insane human or a sane vampire. Reliable people are those who make true statements; unreliable people are those who make false statements (whether out of malice or delusion).

Suppose you ask a Transylvanian: "Are you reliable?" and he gives you a "Yes" or "No" answer. Can you determine from his answer whether or not he is a vampire? Can you determine whether he is sane?

178. _____

Suppose, instead, you asked him, "Do you believe that you are reliable?" He gives you a "Yes" or "No" answer. Now can you determine whether he is a vampire? Can you determine whether he is sane?

B. IS COUNT DRACULA STILL ALIVE?

179. _____

We recall that the first important question I wanted to settle was whether Count Dracula was still alive. Well, I asked one Transylvanian about the matter, and he said, "If I am human, then Count Dracula is still alive."

Can it be determined if Dracula is still alive?

180. _____

Another Transylvanian said, "If I am sane, then Count Dracula is still alive."

Can it be determined if Dracula is still alive?

181. _____

Another one said, "If I am a sane human, then Count Dracula is still alive."

Can it be determined whether Dracula is alive?

182. _____

Suppose a Transylvanian said, "If I am either a sane human or an insane vampire, then Count Dracula is still alive."

Could it then be determined whether Dracula is still alive?

183. _____

Is there a single statement a Transylvanian could make which would convince you that Dracula is alive and also that the statement is false?

184. _____

Is there a single statement a Transylvanian could make which would convince you that Dracula is still alive and which also is such that you could not tell whether the statement is true or false?

185. _____

Suppose a Transylvanian made the following two statements:

(1) I am sane.
(2) I believe that Count Dracula is dead.

Could it be inferred whether Dracula is alive?

186.

Suppose a Transylvanian made the following two statements:

(1) I am human.
(2) If I am human then Count Dracula is still alive.

Could it be determined whether Dracula is still alive?

C. WHAT QUESTION SHOULD BE ASKED?

187.

Can you in one question find out from a Transylvanian whether or not he is a vampire?

188.

Can you in one question find out from a Transylvanian whether or not he is sane?

189.

What question could you ask a Transylvanian which will force him to answer "Yes," regardless of which of the four types he is?

190.

Can you in one question find out from a Transylvanian whether Count Dracula is still alive?

D. IN DRACULA'S CASTLE

Had I had my wits about me and realized the answer to the last problem, I would have saved myself no end of trouble.

But I was so confused at the time, so bewildered by this cross-classification of sane and insane superimposed on lying and truth-telling, that I just could not think straight. Besides, I was a *little* nervous being in the company of Transylvanians, some of whom were vampires. And yet—a far more bewildering situation awaited me!

I still did not know whether Count Dracula was alive. I felt that if only I could get to Dracula's Castle, I could find out the answer. Little did I realize at the time that this would only complicate matters—for reasons you will soon discover.

I knew where Dracula's Castle was all right, and I knew that there was much activity there. I also knew that the castle had a host, but I did not know whether this host was Count Dracula (let alone whether Dracula was even alive!). Now, admission to Dracula's Castle was by invitation only, and invitations were given to only the most elite of Transylvanian society. Therefore, I had to spend several months of arduous social-climbing before I found myself of sufficiently high standing to be invited. The day finally came, and I received an invitation to attend a fête lasting several days and nights at Castle Dracula.

I went with high hopes, and soon received my first shock. A short time after I entered the castle, I realized that in my haste I had forgotten to take my toothbrush, a pocket chess set, and some reading material. So I started to walk out of the door to go back to my hotel, but was intercepted by an exceedingly strong and brutal-looking Transylvanian who politely but quite firmly told me that once one enters Dracula's Castle, he can never leave without permission of the host. "Then," I said, "I should like to meet the host." "That is quite impossible for the present," he informed me, "but I can take a message to him, if you like." Well, I sent the host a written message asking if I could leave the castle for a short while. The reply soon came; it was short and none too reassuring. It said: "Of course not!"

So, here I was a prisoner in the castle of Count Dracula! Well, what could I do? Obviously nothing at the moment, so in a truly Zen-like manner I decided to enjoy the evening for what it was worth and to spring into action whenever the first opportunity presented itself.

The ball that evening was the most magnificent I have ever seen or read about. At about 2:00 A.M. I decided to retire and was shown to my room. Amazingly enough, despite the infinite danger I was in, I slept soundly. I arose about noon the next day, and after a hearty meal I mingled with the guests, hoping to gain more information. Then I received my second shock. All of the people (except myself) belonged to a small, elite subgroup of Transylvanians who instead of using the words "Yes" and "No," used "Bal" and "Da"—just like on the island of zombies! So here I was stuck in a situation with so-called "elite Transylvanians," each of whom was either a human or a vampire, either sane or insane, and on top of all that, I did not know what "Bal" and "Da" meant! Thus the complexities of the former "nonelite" Transylvanians whom I had interrogated outside the castle was compounded with the complexities of zombie island. It seemed that in my coming to the castle I had jumped from the frying pan into the fire.

Well, at this realization, I'm afraid I lost all my Zen-like composure and was thoroughly depressed the rest of the day. I retired early, not even caring to see the second evening of festivities. I lay down wearily, unable either to sleep or to think, Then, suddenly, I jumped up with a start. I realized that the new Bal-Da complications were really easily manageable. I excitedly got out my pencil and notebook and at once worked out the following problems:

191.

In one question (answerable by "Bal" or "Da") I could find out from anyone in the castle whether or not he is a vampire.

192. _____

In one question I could find out if he is sane.

193. _____

In one question I could find out what "Bal" means.

194. _____

If desired, I could ask anyone in the castle a question which would force him to answer "Bal."

195. _____

In one question I could find out whether Dracula is alive!

What are these questions?

E. THE RIDDLE OF DRACULA

Now we come to the climax! Next day I found out all the information I wanted—Dracula was indeed alive, in excellent health, and was in fact my host. To my surprise, I also found out that Dracula was an insane vampire, hence every statement he made was true.

But what good did this knowledge do me now that I was at the mercy of fate and risked being turned into a vampire and losing my soul forever? After a few more days the festivities ended, and all the guests were permitted to leave except for me. So here I was, virtually alone in what was now a dreary macabre castle, a prisoner of a host I had not yet met.

I didn't have long to wait. Shortly before midnight I was rudely awakened from a sound sleep and politely but firmly escorted to the private chambers of Count Dracula, who evidently had requested an audience with me. My guide left, and there I was face to face with Count Dracula himself. After what seemed an eternity of silence, Dracula said, "Are you aware that I always give my victims *some* chance of escape?"

"No," I honestly replied, "I was not aware of this."

"Oh, indeed," replied Dracula, "I would not think of depriving myself of this great pleasure."

Somehow or other, I did not quite like the tone of voice in which he said this; it somehow savored of the supercilious.

"You see," continued Dracula, "I ask my victim a riddle. If he correctly guesses the answer within a quarter of an hour, I set him free. If he fails to guess, or if he guesses falsely, I strike, and he becomes a vampire forever."

"A sane or an insane one?" I innocently inquired.

Dracula turned livid with rage. "Your jokes are not funny!" he shouted. "Do you fully realize the gravity of the situation? I am hardly in the mood for frivolous jests. Any more of that, and I won't even give you the usual chance."

Frightening as all this sounded, my immediate reaction was primarily curiosity as to *why* Dracula would willingly risk losing a victim. "What motivates you to this sporting generosity?" I inquired.

"Generosity?" said Dracula with a disdainful air. "Why, I don't have a generous bone in my body. It's just that the enormous sadistic pleasure I derive in watching my victim squirm, write, and wriggle under these agonizing mental gymnastics more than compensates for the infinitesimal probability that I will lose him."

This word "infinitesimal" was none too consoling. "Oh yes," continued Dracula, "I have never lost a victim yet; so you see, I am not running much risk."

"Very well," I said, bracing myself as well as I could, "what is the riddle?"

196. _____

Dracula looked at me scrutinizingly for some time. "Your questions to my guests were very clever—oh yes, I know all about them. Very clever indeed, but not as clever as you might think. You had to design a separate question for each piece of information you wanted; you never hit on one simple unifying principle which would have saved you much mental labor. There is one sentence S having the almost magical property that given any information you want to know, given *any* sentence X whose truth you wish to ascertain, all you would have to do is ask anyone in this castle, 'Is S equivalent to X?' If you get 'Bal' for an answer, X must be true; if you get 'Da' for an answer, X must be false. So, for example, if you wished to find out whether the speaker is a vampire, you would ask, 'Is S true if and only if you are a vampire?' If you wished to find out if he is sane, you need merely ask, 'Is S true if and only if you are sane?' To have found out what 'Bal' means, you needed merely to ask, 'Is S true if and only if "Bal" means *yes*?' To have found out whether I was still alive, you could have asked, 'Is S true if and only if Dracula is still alive?' etc."

"What is this sentence S?" I asked, with enormous curiosity. "Ah" replied Dracula, "that is for *you* to find out! This is your riddle!"

So saying, Dracula rose to leave the room. "You have fifteen minutes. You'd better think hard; the stakes are quite high."

Quite high, indeed! Those were the most painful fifteen minutes of my life. I was so paralyzed by fear that no thoughts came at all. I felt certain that Dracula was secretly watching me from some hiding place.

When fifteen minutes elapsed, Dracula triumphantly returned and started lumbering toward me with dripping fangs. Closer and closer he came until he was practically upon me. Then suddenly I raised my hand and yelled: "Of course! The sentence S is. . . ."

What is the sentence S which saved me?

Epilogue. ———————————————————————

The shock on poor Dracula on my having solved the riddle

was so great that he perished on the spot and, within a few minutes, crumbled into dust. Now when anyone asks me, "Is Count Dracula still alive?" I can truthfully and accurately answer "Bal."

197. _____

There are four minor inconsistencies in this story. Can you spot them?

SOLUTIONS _____

167. _____

His statement is either true or false. Suppose it is false. Then he is neither human nor sane, hence he must be an insane vampire. But insane vampires make only true statements, and we have a contradiction. Therefore his statement is true. The only ones who make true statements are sane humans or insane vampires. If he were an insane vampire then he wouldn't be either human or sane, and his statement would be false. But we know the statement is true. Therefore he must be a sane human.

168. _____

He must be an insane vampire.

169. _____

No, this time he is a sane vampire.

170. _____

A sane human would answer "No" to this question, and any of the other three types would answer "Yes." Had I gotten a "Yes" answer, I couldn't have known what type he was. But I told you that I did know, hence he didn't answer "Yes." So he answered "No," from which follows that he must have been a sane human.

171. _____

It cannot be inferred whether he is human or a vampire, but it does follow that he is insane. A sane human would not say that he is a vampire and a sane vampire would know that he is a vampire and would then lie and say he is human. On the other hand, an insane human would believe, and hence would say he is a vampire, and an insane vampire would believe he was human and would then say he is a vampire.

172. _____

This time, all that follows is that he is a vampire. A sane human could not say that he is insane, and an insane human would believe that he is sane, and, being human, could not say that he is insane.

173. _____

I'm sure many such pairs of statements can be found; the pair I had in mind is this:

> X: If I am sane, then I am human.
> Y: If I am human, then I am sane.

Suppose the speaker asserts X. We will prove that Y must be true, that is, that if he is human, then he is sane. Well, suppose he is human. Then it is true that if he is sane then he is human (since he is human, period). This means that X is true. Then the speaker must be sane, because insane humans don't make true statements. Therefore if he is human, he is sane, hence Y is true.

Conversely, suppose the speaker asserts Y. We must show X true. Well, suppose he is sane. Then Y must be true. Hence the speaker is human (because sane vampires don't make true statements). So he is human (under the assumption that he is sane). Therefore if he is sane then he is human, so statement X is true.

174. _____

The answer to both questions is "Yes." Suppose a Transylvanian believes a certain statement X. Then it of course does not follow that X must be true, because he may be insane. But if he believes that he believes X, then X must be true! For, suppose on the one hand that he is sane. Since he believes the statement that he believes X, then the statement that he believes X must be true. Therefore he in fact does believe X. And since he is sane, X must be true. On the other hand, suppose he is insane. Since he believes the statement that he believes X, then the statement that he believes X must be false. Hence he doesn't really believe X (he only thinks he does!). Since he doesn't believe X, and he is insane, then again X must be true.

We have thus shown that if a Transylvanian believes that he believes X, then X must be true regardless of whether he is sane or insane. Similarly it can be shown that if he doesn't believe that he believes X, then X must be false. We leave this to the reader.

175. _____

Again both answers are "Yes"—this is a corollary of the solution to the preceding problem.

Suppose A asserts that he believes X. Suppose A is human. Then he believes what he asserts, so he believes that he believes X. Then, as we have seen in the solution to problem is 174, X must be true, whether A is sane or insane. Similarly, suppose A is a vampire. Then he doesn't believe what he asserts, so he doesn't believe that he believes X. So X must be false, whether A is sane or insane.

176.

A asserts that he believes that B is human. B either asserts that he believes A is human or asserts that he believes that A is not human. If the latter were the case, we would get the following contradiction.

We have:

(1) A says that he believes B is human.
(2) B says that he believes A is not human.

Suppose A is human. Then by (1) it follows, by the principle of problem 175, that B is human. Then by (2) it follows (by the same principle) that A is not human. Therefore it is a contradiction that A is human.

Suppose A is a vampire. Then from (1), B is not human (by the same principle), so B is a vampire. Then from (2) it follows (by the same principle) that A is human. This is again a contradiction. Therefore if B answered "No" we would have a contradiction. Hence B answered "Yes."

177.

Nothing whatever can be inferred, because all Transylvanians will answer "Yes" to this question. The reader can check this out for himself.

178.

This is a different case; it cannot be inferred from the answer whether the speaker is human or a vampire, but it can be inferred whether he is sane. If he is sane, then he will answer "Yes"; if he is insane then he will answer "No." We leave the proof to the reader.

179.

No, it cannot. It could be that he is a sane human and

Dracula is alive, or it could be that he is an insane vampire and Dracula is dead. (In fact, if he is an insane vampire then Dracula could be alive or dead.)

180. _____

Again the answer is "No."

181. _____

The answer is still "No." He could, for example, be an insane vampire, in which case Dracula might or might not be alive.

182. _____

Yes, this time it would follow that Dracula is alive.

Let us use the terminology of problem 177 and re-phrase the native's statement thus: "If I am reliable then Dracula is alive."

We proved in Chapter 8 (see solutions to problems 109–112) that if a native of an island of knights and knaves says, "If I am a knight then so-and-so," then the speaker must be a knight and the so-and-so must be true. Similarly, if an inhabitant of Transylvania says, "If I am reliable then so-and-so," then he must be reliable and the so-and-so must be true. The proof is really the same—just substitute the word "reliable" for "a knight."

183. _____

A statement which would work is: "I am unreliable and Dracula is dead." We leave the proof to the reader. (Hint: first show that the speaker is not reliable.)

184. _____

A sentence which does this is: "I am reliable if and only if Dracula is still alive."

In the solution to problem 122 of Chapter 8 we proved that if an inhabitant of an island of knights and knaves says, "I am a knight if and only if so-and-so," then the so-and-so must be true (but it is not possible to tell whether the speaker is a knight or a knave). Similarly, if a Transylvanian says, "I am reliable if and only if so-and-so," then the so-and-so must be true regardless of whether the speaker is reliable or not. The proof is really the same—just substitute "reliable" for "a knight."

There are several other statements which would also work. For example: "I believe that the statement that Dracula is alive is equivalent to the statement that I am human." Another, rather amusing, example is "I believe that if someone asked me whether Dracula is still alive, then I would answer "Yes."

185. _____

Yes, it would follow that Dracula must be dead.

From (1) we can infer that the speaker is human because a sane vampire would know he is sane and hence say he is insane, and an insane vampire would believe he is sane and then say that he is insane. Therefore the speaker is human.

Let us now recall the principle established in problem 175: when a human says that he believes something, then that something must be the case (regardless of whether he is sane or insane). Well, we now know the speaker is human and that he said that he believes that Dracula is dead. Therefore Count Dracula must be dead.

186. _____

From his first statement, "I am human," it follows, not that he is human, but that he must be sane. (An insane human wouldn't know he was human and an insane vampire would think he is human and then would say he was a vampire.) Now that we know that he is sane, we shall prove that he is human. Suppose he were a vampire. Then it is false that he is human, and since a false statement implies any statement, then his second statement—"If I am human then Dracula is still alive"—would have to be true.

But a sane vampire cannot make true statements, so we have a contradiction. Therefore he cannot be a vampire; he must be human.

Now we know that he is both sane and human, so he makes true statements. Therefore his second statement—that if he is human then Dracula is still alive—must be true. Also, he is human. Therefore Dracula is still alive.

187.

Just ask him whether he is sane. A human (whether he is sane or not) will answer "Yes" and a vampire will answer "No."

188.

Just ask him whether he is a human. A sane Transylvanian (whether he is human or vampire) will say "Yes," and an insane Transylvanian will say "No."

For the next few problems I will just tell you what the questions are. You should have enough experience by now to be able to prove for yourselves that these questions work.

189.

One question which works is: "Do you believe you are human?" All Transylvanians must answer "Yes" to this question. It's not that they all believe that they are human (only sane humans and insane vampires believe this) but all natives will say that they believe it.

Another question which would work is: "Are you reliable?" All Transylvanians would claim to be reliable.

190.

Either of the following questions will work:

    (1) "Is the statement that you are reliable equivalent to the statement that Dracula is alive?

    (2) "Do you believe that the statement that you are a human is equivalent to the statement that Dracula is alive?"

191.

Ask him, "Is 'Bal' the correct answer to the question of whether you are sane?" If he answers "Bal" then he is human; if he answers "Da" then he is a vampire.

192.

Ask him, "Is 'Bal' the correct answer to the question of whether you are human?" If he answers "Bal" then he is sane; if he answers "Da" then he is insane.

193.

Ask him, "Do you believe you are human?" Whatever word he answers must mean *yes*. Alternatively, ask him, "Are you reliable?"

194.

One question which will work is: "Is 'Bal' the correct answer to the question of whether you are reliable?" (We recall that being reliable means being either a sane human or an insane vampire.)

    Another question which works: "Are you reliable if and only if 'Bal' means *yes*?"

    Either of these questions will force an answer of "Bal," as can be proved in essentially the same manner as problem 161 of Chapter 11 (except that being reliable now plays the role played by being human).

195. _____

Either of the following questions will do the job.

(1) Do you believe that "Bal" is the correct answer to the question of whether the statement that you are human is equivalent to the statement that Dracula is alive?
(2) Is "Bal" the correct answer to the question of whether the statement that you are reliable is equivalent to the statement that Dracula is alive?

A much simpler and more elegant solution is provided by the unifying principle, which is explained in number 196.

196. The Unifying Principle. _____

Let us define an elite Transylvanian to be of type 1 if he answers "Bal" to the question: "Does 2 plus 2 equal 4?" This means, of course, that given any other question whose correct answer is "Yes," one of type 1 will answer "Bal" to this question. We will define an elite Transylvanian to be of type 2 if he is not of type 1. This means that given any true statement X (such as 2 plus 2 equals 4), if you ask one of type 2 whether X is true, he will answer "Da."

Let us immediately note that if "Bal" does mean *yes*, then people of type 1 are those who are reliable, and people of type 2 are those who are unreliable. If "Bal" means *no*, then we have the reverse (type 1=unreliable and type 2=reliable).

Now, the unifying principle is this: To find out of any given statement X whether X is true, just ask an elite Transylvanian whether X is equivalent to the statement that he is of type 1. You could phrase your question thus: "Is X true if and only if you are of type 1?" We will prove that if he answers "Bal," then X must be true, and if he answers "Da," then X must be false. Thus the "magic" sentence S is: "You are of type 1" (or "You answer 'Bal' to the question of whether $2+2=4$").

Proof: S is the sentence: "You are of type 1"; X is the sentence whose truth you wish to ascertain. The question you ask is whether S is equivalent to X. Suppose you get the answer "Bal." We are to prove that X must then be true.

Case One: "Bal" means yes. In this case we know two things: (i) type 1=reliable; (ii) the speaker, by saying, "Bal," is asserting that S is equivalent to X.

Subcase 1a: The speaker is of type 1. Then he is reliable and makes true statements. Then S really is equivalent to X and also S is true (since he is of type 1). Hence X is true.

Subcase 1b: The speaker is of type 2. Then he is unreliable and makes false statements. Since he asserts that S is equivalent to X, then S is not equivalent to X. But S is false (since the speaker is not of type 1), and X is not equivalent to S, so X is true.

Case Two: "Bal" means no. In this case we know two things: (i) type 1=unreliable; (ii) the speaker is asserting that S is not equivalent to X.

Subcase 2a: The speaker is of type 1. Then he is unreliable and makes false statements. He falsely asserts that S is not equivalent to X, hence S really is eqivalent to X since S is true, then X is true.

Subcase 2b: The speaker is of type 2. Then he is reliable and makes true statements. Hence S is not equivalent to X (since he asserts it isn't), but S is false, hence X must again be true.

We have shown that a "Bal" answer means that X is true. We could go through a similar round of reasoning to prove that a "Da" answer signifies that X is false. However, we can take the following shortcut:

Suppose he answers "Da." Now, answering "Da" to this question is really the same as answering "Bal" to the question: Are you of type 1 if and only if X is false?" (Because for any two statements Y and Z, the statement that Y is equivalent to Z is the very opposite of the statement that Y is equivalent to not Z). So he would have answered "Bal" had you asked him: "Are you of type 1 if and only if X is false?" Since he would have answered "Bal" to this, then it follows (by the above proof) that X really is false.

197. Answer to the Question on Inconsistencies.

(1), (2) On two occasions Dracula said, "Oh yes." An elite Transylvanian does not use the word "Yes."

(3) When the strong and brutal-looking Transylvanian told me that I could not leave the castle without permission of the host, why should I have believed him?

(4) When the host sent me back the message "Of course not!" why should I have believed him? I did not yet know that the host was an insane vampire and makes correct statements.

PART FOUR

Logic Is a Many-Splendored Thing

13. Logic and Life

A. SOME CHARACTERIZATIONS OF LOGIC

198. Tweedledum's Characterization of Logic. ___
I love the following characterization of logic given by Tweedledum:

> *Tweedledee* (to Alice) / I know what you're thinking about, but it isn't so, nohow.
> *Tweedledum* / Contrariwise, if it was so, it might be; and if it were so, it would be; but as it isn't it ain't. That's logic.

199. Thurber's Characterization. ___
In *The Thirteen Clocks,* Thurber gives a characterization of logic which goes something like this: Since it is possible to touch a clock without stopping it, then it is possible to start a clock without touching it. This is logic as I see and understand it.

200. ___
Thurber's characterization reminds me a bit of my

favorite syllogism: Some cars rattle. My car is some car. So no wonder my car rattles!

201. Another Characterization of Logic. ___

A friend of mine—an ex-police officer—when he heard I was a logician, said: "Let me tell you how I see logic. The other day my wife and I were at a party. The hostess offered us some cake. On the platter were just two pieces, one larger than the other. I thought for a while, and then I decided to take the larger piece. Here is how I reasoned: I know my wife likes cake and I know she knows that I like cake. I also know she loves me and wants me to be happy, therefore she would want me to have the larger piece. Therefore I took the larger piece."

202. ___

The above reminds me of the story of two men who were in a restaurant and ordered fish. The waiter brought a dish with two fish, one larger than the other. One of the men said to the other, "Please help yourself." The other one said, "Okay," and helped himself to the larger fish. After a tense silence, the first one said, "Really, now, if you had offered me first choice, I would have taken the smaller fish!" The other one replied, "What are you complaining for; you have it, don't you?"

203. ___

This also reminds me of the story of a woman at a banquet. When the silver platter of asparagus came her way, she cut off all the tips, put them on her plate and passed the platter to her neighbor. The neighbor said: "Why do you do a thing like that? Why do you keep all the tips for yourself and pass the rest on to me?" The woman replied, "Oh, the tips are the best part, didn't you know?"

204.

I once saw this cartoon in a newspaper: A little boy and girl are walking on a sidewalk; the boy is walking on the inside. A truck has just passed through the muddy street and splashed the girl hopelessly. The boy says, "Now do you understand why I don't walk on the outside like a gentleman?

205.

I also like the following characterization of ethics. A boy once asked his father, "Daddy, what is ethics?" The father replied: "I will explain it to you, my son. The other day a lady came into the store. She gave me a twenty dollar bill, thinking it was a ten. I also thought it was a ten, and gave her change accordingly. Several hours later I discovered it was a twenty. Now ethics, my boy, is: 'Should I tell my partner?' "

206.

I once went into a Chinese restaurant with a mathematician friend. On the menu was printed: *Extra charge for anything served extra*. My friend observed, "They really could have left out the first and last words."

207.

I once saw the following sign outside a restaurant.

> GOOD FOOD IS NOT CHEAP
> CHEAP FOOD IS NOT GOOD

Do these two sentences say the same thing or different things?

The answer is that logically speaking, they say exactly the same thing; they are both equivalent to the statement that no food is both good and cheap. Though these statements are *logically* equivalent, I would say that *psychologically* they suggest different things; when I read the first sentence, I picture good expensive food; when I read the second, I think of cheap rotten food. I don't think my reaction is atypical.

B. ARE YOU A PHYSICIST OR A MATHEMATICIAN?

208.

A well-known problem concerns two beakers, one containing 10 fluid ounces of water and the other 10 fluid ounces of wine. Three ounces of the water are poured into the wine container, and after stirring, 3 ounces of the mixture are poured back into the water container. Is there now more water in the wine container or more wine in the water container?

There are two ways of solving this problem, one by straightforward arithmetic and the other by common sense. Of the two, I greatly prefer the latter. The solution by arithmetic is as follows: After 3 ounces of water are poured into the wine container, there are then 13 ounces of mixture in the wine container; so the mixture is 3/13 water and 10/13 wine. After I pour 3 ounces of the mixture back into the water container, I have poured $3 \times 10/13 = 30/13$ ounces of wine into the water. So the water container now contains 30/13 ounces of wine. Now, before the second pouring, the wine container contained 3 ounces of water, and $3 \times 3/13$ ounces of water was poured back into the water container. So the wine container now contains $3 - 9/13$ ounces of water. But $3 - 9/13 = 39/13 - 9/13 = 30/13$. So the wine container contains exactly the same amount (*viz.*, 30/13) of water as the water container contains wine.

The common-sense solution is far quicker, and also

suggests something far more general: Since the amount of liquid in each container is now the same, then obviously however much water is missing from the water container is replaced by the same volume of wine. This solves the problem. Of course this "common-sense" solution doesn't tell you what this volume is, whereas the arithmetic solution tells you that it is 30/13. However, the common-sense solution is equally applicable to the following more general problem (which the arithmetic method could never handle).

We start with the same containers as before, and we pour liquid back and forth from one container to the other, without specifying how much we pour or how many pourings we make, nor is it necessary that the same amount be poured each time, but when we are finished, we have 10 ounces of liquid in each container. Is there more water in the wine container or more wine in the water container?

By the same common-sense argument, the amounts must be equal, but there is now no way of knowing what this amount is.

209. _____

When I came upon the above problem, I immediately thought of the following question: We start again with the 10 ounces of water in the first beaker, A, and 10 ounces of wine in the second beaker, B. We transfer 3 ounces back and forth any finite number of times. What is the smallest number of pourings required to reach a stage at which the percentage of wine in each mixture is the same?

The solution I had in mind is that it is impossible to do this in any finite number of steps. Regardless of how much wine is in one beaker and how much water is in the other, and regardless of how much liquid is poured back and forth at each step (provided one never empties one beaker into the other), the wine concentration in B will always be higher than that in A. This can be shown by a simple mathematical induction argument. At the outset, the wine concentration in B is of course higher than in A. Now, suppose after a given stage the B is still more concentrated than A. If we pour some of B into A, we are pouring from a stronger to a weaker mixture, hence B will still be stronger than A. If we pour from A to B, then B will still be stronger than A. Since every transfer is one of these two cases, it follows that B must always remain more concentrated than A. The only way to equalize the mixture is by pouring all of one beaker into the other.

Now, as a purely *mathematical* problem, my reasoning is impeccable. As a problem about the actual physical world, however, my reasoning was quite fallacious. It assumed that liquids are infinitely divisible, whereas in fact they are composed of discrete molecules. This was pointed out to Martin Gardner by P. E. Argyle of Royal Oak, British Columbia. Argyle calculated that after 47 double interchanges, the probability would be significant that the concentrations would be the same.[1]

I wonder if Argyle's solution is correct if the number of molecules in the wine container is odd rather than even. At any rate, I never in a million years would have thought of this as a physical rather than a mathematical problem.

210. Magnet Testing. _____

Martin Gardner gives the following problem:[2] You are in a room containing no metal of any sort except for two iron bars. One is a bar magnet and the other is not magnetized. You can tell which one is the magnet by suspending each by a thread tied around its center and observing which bar tends to point north. Is there a simpler way?

The given solution was to pick up one of the bars and touch its end to the middle of the other bar. If there is magnetic attraction, then you are holding the magnet; if there isn't, then you are not.

[1]For details see Martin Gardner, *The Second Scientific American Book of Puzzles and Diversions* (New York: Simon and Schuster, 1961), pp. 163-64.
[2]Martin Gardner, *Mathematical Carnival* (New York: Vintage Books, 1977), p. 178.

This "physicist's" solution is a perfectly sensible one and is certainly simpler than the expedient of suspending both bars by threads tied around their centers. Well, I, being essentially a logician rather than a physicist, thought of the following solution, which I believe is midway in simplicity between the two others: namely, suspend just *one* magnet by a thread tied to its center and see if *it* points north.

211. And What About You?

Are *you* of the mathematician or physicist type? Well, there is the following delightful test to tell whether you are a mathematician or physicist.

You are in a country cabin in which there is an unlighted stove, a box of matches, a faucet with cold running water, and an empty pot. How would you get a pot of hot water? Doubtless you will answer, "I would fill the pot with cold water, light the stove, and then put the pot on until the water gets hot." To this I reply: "Good; so far, mathematicians and physicists are in complete agreement. Now, the next problem separates the cases."

In this problem, you are in a country cabin in which there is an unlighted stove, a box of matches, a faucet with cold running water, and a pot filled with cold water. How would you get a pot of hot water? Most people reply, "I would light the stove and put the pot of cold water on it." I reply: "Then you are a physicist! The mathematician would pour out the water, reducing the case to the preceding problem, which has already been solved."

We could go a step further and consider the case of a pot of cold water already on a lighted stove. How do we get hot water? The physicist just waits for the water to get hot; the mathematician turns off the stove, dumps out the water, reducing the case to the first problem (or he might just turn off the stove, reducing the case to the second problem).

A still more dramatic variation goes as follows: A

house is on fire. We have available a hydrant and a disconnected hose. How does one put out the fire? Obviously, by first connecting the hose to the hydrant and then squirting the building. Now, suppose you have a hydrant, a disconnected hose and a house not on fire. How do you put out the fire? The mathematician first sets fire to the house, reducing the problem to the preceding case.

212. Von Neumann and the Fly Problem. _____

The following problem can be solved either by the "hard" way or the "easy" way.

Two trains 200 miles apart are moving toward each other; each one is going at a speed of 50 miles per hour. A fly starting on the front end of one of them flies back and forth between them at a rate of 75 miles an hour. It does this until the trains collide and crush the fly to death. What is the total distance the fly has flown?

The fly actually hits each train an infinite number of times before its gets crushed, and one could solve the problem of summing an infinite series of distances (getting shorter and shorter, of course, and converging to a definite finite amount)—this is solving it the "hard" way and would have to be done with pencil and paper. The "easy" way is as follows: Since the trains are 200 miles apart and each train is going at 50 miles an hour, it takes 2 hours for the trains to collide. Therefore the fly was flying for 2 hours. Since the fly was flying at the rate of 75 miles per hour, then the fly must have flown 150 miles. That's all there is to it!

Well, the great mathematician Von Neumann was given this problem, thought for a few seconds and said, "Oh, of course, 150 miles." His friend said, "Good, how did you get it?" Von Neumann replied, "I summed the series."

213. _____

There is also the following joke about Von Neumann. He was consulted by a group who was building a rocket ship to send into outer space. When he saw the incomplete structure, he asked, "Where did you get the plans for this ship?" He was told, "We have our own staff of engineers." He disdainfully replied: "Engineers! Why I have completely sewn up the whole mathematical theory of rocketry. See my paper of 1952." Well, the group consulted the 1952 paper, completely scrapped their 10 million dollar structure, and rebuilt the rocket exactly according to Von Neumann's plans. The minute they launched it, the entire structure blew up. They angrily called Von Neumann back and said: "We followed your instructions to the letter. Yet when we started it, it blew up! Why?" Von Neumann replied, "Ah yes; that is technically known as the *blow-up problem*—I treated that in my paper of 1954."

214. _____

There is an allegedly true story about a little girl living in Princeton, New Jersey, who was having trouble with arithmetic. In a period of about two months she, for some unknown reason, made a startling improvement. One day her mother asked her if she knew the reason for her improvement. The little girl replied: "I heard there is a professor in this town who's good at numbers. I rang his doorbell, and every day he's been helping me. He teaches real good." The mother, somewhat startled, asked her whether she knew his name. The little girl replied, "Not exactly; it goes something like *Ein-stein.*"

215. _____

There is another story that Einstein once told a colleague that he did not like teaching at a co-ed college because with all the pretty girls in the room, the boys wouldn't pay enough attention to mathematics and physics. His friends said, "Oh come on now, Albert, you know the boys would listen to what *you* have you say." Einstein replied, "Oh, such boys are not worth teaching."

216. _____

The following joke illustrates perfectly the difference between a physicist and a mathematician.

A physicist and a mathematician were flying together from the West Coast to a research laboratory in Washington, D.C. Each was asked to write a report of his trip. Well, over Kansas they passed a black sheep. The physicist wrote: "There is a black sheep in Kansas." The mathematician wrote: "There exists—somewhere in the Midwest—a sheep—black on top."

C. VERMONTERS

217. _____

The last story is reminiscent of a story told about Calvin Coolidge. Coolidge was visiting a farm with some friends. When they came to a flock of sheep, one of the friends said, "I see these sheep have just been shorn." Coolidge replied, "Looks like it from this side."

218. _____

When the humorist Will Rogers was about to be introduced to President Coolidge, he was told, "You know, it's impossible to make Coolidge laugh." Rogers said, "I'll make him laugh." And Will Rogers most certainly did! When introduced to the president, and when told, "Mr. Rogers, I would like you to meet President Coolidge," Will Rogers turned to the president and said: "Eh? Didn't get the name."

219. _____

Calvin Coolidge was, of course, a Vermonter, and I love stories about Vermonters. One story goes that a man walked past the house of a Vermont farmer who was sitting on the porch rocking in his chair. The man said, "Been rocking like that all your life?" The farmer replied, "Not yet!"

220. _____

A characteristic of Vermonters (at least as portrayed in humorous stories) is that the Vermonter, when asked a question, gives accurate answers but often fails to include information which may be highly relevant and very important. A perfect illustration of this principle is the joke about one Vermont farmer who went to his neighbor's farm and asked the other farmer, "Lem, what did you give your horse last year when it had the colic?" Lem replied, "Bran and molasses." The farmer went home, returned one week later, and said, "Lem, I gave my horse bran and molasses, and it died." Lem replied, "So did mine."

221. _____

My favorite Vermonter story is about the tourist traveling in Vermont who came across a fork in the road. On one road was a sign: "To White River Junction." On the other road was a sign: "To White River Junction." The tourist scratched his head in perplexity, spied a Vermont native standing at the intersection, went over to him and asked, "Does it make any difference *which* road I take?" The Vermonter replied, "Not to me it doesn't."

D. OBVIOUS? _____

222. _____

This story has been told about many different mathematicians. A mathematics professor during a lecture made a statement and then said, "This is obvious." A student raised his hand and asked, "Why is it obvious?" The professor thought for a few moments, walked out of the room, returned about twenty minutes later, and said, "Yes, it is obvious!"—and continued the lecture.

223. _____

Another story is told about a professor who met a student in the hall shortly after he had given a lecture. The student said: "Professor——, I did not understand the proof you gave of Theorem 2. Could you please explain it again?" The professor went into a trance-like silence for about three minutes, and then said, "Yes, therefore it follows!" The student replied, "But what is the proof?" The professor went into another trance, returned to earth, and said, "—therefore the proof is correct." The student replied, "Yes, but you still haven't told me what the proof *is*!" The professor said, "All right, I'll prove it to you another way!" He went into another trance, returned, and said, "That also does it." The poor student was as bewildered as ever. The professor said, "Look, I've given you three proofs; if none of these help, I'm afraid there is nothing more I can do," and walked away.

224. _____

A story is told about a famous physicist who, after a lecture to a professional group, said, "Now I will take any questions." One member of the audience raised his hand and said, "I didn't understand your proof of Theorem B." The physicist replied, "That's not a question."

225. _____

When I was a graduate student at Princeton, there was circulating the following explanation of the meaning of the word "obvious" when used by different members of the mathematics department. I shall not use names, but letters.

When Professor A. says something is obvious, it means that if you go home and think about it for a couple of weeks, you will realize it is true.

When Professor L. says something is obvious, it means that if you go home and think about it for the rest of your life, the day *might* come when you will see it.

When Professor C. says something is obvious, it means that the class has already known it for the last two weeks.

When Professor F. says something is obvious, it means that it is probably false.

E. ABSENT-MINDED PROFESSORS

226.

One story has it that a student one day met a professor in the hall. He asked him, "Have you had lunch yet?" The professor thought for a moment and said, "Tell me, in which direction was I walking when you stopped me?"

227.

I heard the following story about the mathematician David Hilbert. I once told this story to a physicist who told me that he had heard that same story about Ampere!

As I heard the story, Professor and Mrs. Hilbert were giving a party. After one guest arrived, Mrs. Hilbert took David aside and said, "David, go up and change your tie." Hilbert went up; an hour passed and he didn't come down. Mrs. Hilbert was worried, went up to the bedroom, and found Hilbert in bed asleep. When awakened, he recalled that when he took off his tie, he automatically went through the motions of taking off the rest of his clothes, putting on his pajamas, and getting into bed.

228.

My favorite of all absent-minded professor stories is one

told about Norbert Weiner. I have no idea whether or not it is true (though it conceivably could be, since Weiner did have very poor eyesight in his later years), but whether true or not, here it is.

The Weiners were to move from one part of Cambridge to another. Mrs. Weiner, knowing of her husband's absent-mindedness, decided to condition him in advance. So thirty days before the moving date Mrs. Weiner said to her husband in the morning before he left for school: "Now Norbert, thirty days from now we will move. When you get out of class, you don't take bus A; you take bus B!" Weiner replied, "Yes, dear." The next morning Mrs. Weiner said: "Now remember, Norbert, in twenty-nine days we will move. When you get out of class, you don't take bus A; you take bus B!" Weiner replied, "Yes, dear." Well, this went on each day until the morning of the moving day. Mrs. Weiner said, "Now *today* is the day, Norbert: when you get out of class today, you don't take bus A; you take bus B!" Norbert replied, "Yes, dear." Well, when Weiner got out of class, he of course took bus A, walked to his house, and found it empty. He said to himself: "Oh, of course! This is the day we have moved!" So he went back to Harvard Square, took bus B, and got off at what he remembered was the correct stop. However, he had forgotten his new address. He wandered around, and by this time it was getting quite dark. He spied a little girl on the street, went over to her, and said, "Excuse me, but would you by any chance happen to know where the Weiners live?" The little girl replied, "Oh, come on, Daddy, I'll take you home."

F. MUSICIANS

229. _____

The composer Robert Schumann wrote at the beginning of one of his compositions: "To be played as fast as possible." A few measures later he wrote: "Faster."

230.

A story is told that Richard Wagner was walking on a street in Berlin one day and came across an organ-grinder who was grinding out the overture to *Tannhauser*. Wagner stopped and said, "As a matter of fact, you are playing it a little too fast." The organ-grinder at once recognized Wagner, tipped his hat, and said, "Oh thank you, Herr Wagner! Thank you, Herr Wagner!"

The next day Wagner returned to the same spot and found the organ-grinder grinding out the overture at the correct tempo. Behind him was a big sign: "PUPIL OF RICHARD WAGNER."

231.

There is the story of four musicians from the Boston Philharmonic who were out rowing. One of them fell overboard and yelled: "Help! I can't swim!" One of the other musicians yelled, "Fake it!"

232. Brahms and the Amateur String Quartet.

This story is told of the composer Johannes Brahms, who had four friends who were string players. They were very poor musicians, but such nice people that Brahms enjoyed associating with them. They decided to surprise Brahms and spent six months assiduously practicing Brahms' latest quartet. One evening they cornered Brahms at a party, and the first violinist said: "Johannes, we have a surprise for you. Come into the next room please." Brahms followed them into the next room, the players took out their instruments and started to play the quartet. Well, the first movement was about as much as poor Brahms could bear! He got up, gave a polite but sickly smile, and started to leave the room. The first violinist ran after him and said: "Johannes, how was the performance? Was the tempo all right?" Brahms replied: "Your tempos were all good. I think I liked *yours* the best."

233. _____

Many experiments have been conducted in which an English sentence (preferably an idiom) is translated by one computer into Russian, then a second computer translates the Russian back to English. The purpose of the experiment is to see how much distortion results.

In one case they tried the idiom: "The spirit is strong but the flesh is weak." What came back was: "The vodka is good, but the meat is rotten."

234. _____

Another time they tried the idiom: "Out of sight, out of mind." What came back was: "Blind idiot."

235. _____

There is the joke of an IBM salesman who tried to sell a computer that "knew everything." The salesman said to one customer, "Ask it anything you like; it will answer you." The customer said, "Okay, where is my father?" The machine thought for a minute, and out came a card which said: "Your father is now fishing in Canada." The customer said: "Ha! The machine is no good! It so happens that my father has been dead for several years." The salesman replied: "No, no; you have to ask in more precise language! Here, let me ask the question for you." He stepped over to the computer and said, "This man before you; where is his mother's husband?" The computer thought for a moment, and out came a card: "His mother's husband has been dead for several years. His father is now fishing in Canada."

236. _____

When the world's first automated plane took off, the passengers were a bit worried. Then the computer's soothing, reassuring voice came over the loudspeaker: "Ladies and gentlemen, you are privileged to be riding the world's first fully automated plane. No human erring pilots; you are guided by infallible computers. All your needs will be taken care of. You have nothing to worry about—worry about—worry about—worry about— ..."

237. The Military Computer. _____

My favorite computer story is about a military computer. The army had just sent a rocket ship to the moon. The colonel programmed two questions into the computer: (1) Will the rocket reach the moon? (2) Will the rocket return to earth? The computer thought for a while, and out came a card which said: "Yes." The colonel was furious; he didn't know whether "Yes" was in answer to the first question or the second question or the conjunction of the two questions. So he angrily programmed back: "Yes, *what?*" The computer thought for a while, and a card came out saying: "Yes, Sir."

14. How to Prove Anything

I think a good characterization of a drunken mathematician is one who says, "I can prove anyshing!"

In Plato's dialogue *Euthydemus,* Socrates, in describing to Crito the amazing dialectical talents of the sophist-brothers Euthydemus and Dionysodorus, says, "So great is their skill that they can refute any proposition whether true or false." Later in the dialogue Socrates describes how Dionysodorus proves to one of the audience, Ctessipus, that Ctessipus' father is a dog. The argument is as follows:

> *Dion* / You say you have a dog?
>
> *Ctes* / Yes, a villain of one.
>
> *Dion* / And has he puppies?
>
> *Ctes* / Yes, and they are very like himself.
>
> *Dion* / And the dog is the father of them?
>
> *Ctes* / Yes, I certainly saw him and the mother of the puppies come together.
>
> *Dion* / And is he not yours?
>
> *Ctes* / To be sure he is.
>
> *Dion* / Then he is a father and he is yours; ergo, he is your father, and the puppies are your brothers.

Inspired by the example of these great sophists, I shall, in this chapter, prove to you many strange and wondrous things.

A. PROOFS OF VARIOUS AND SUNDRY THINGS

238. Proof that Either Tweedledum or Tweedledee Exists.

This proof will not show that Tweedledee and Tweedledum *both* exist; it will merely show that at least one of them does. Moreover, it will be impossible to tell from the proof which one of them really exists.

We have a box in which is written the following three sentences:

```
(1) TWEEDLEDUM DOES NOT EXIST
(2) TWEEDLEDEE DOES NOT EXIST
(3) AT LEAST ONE SENTENCE IN THIS BOX IS FALSE
```

Consider sentence (3). If it is false, then it is not the case that at least one of the three sentences is false, which means that all three sentences are true, which means that sentence (3) is true, and this is a contradiction. Therefore sentence (3) cannot be false; it must be true. Hence at least one of the three sentences really is false, but it can't be (3) that is false, hence sentence (1) or sentence (2) is false. If sentence (1) is false, then Tweedledum exists; if sentence (2) is false then Tweedledee exists. Hence either Tweedledum or Tweedledee exists.

I once gave a talk on my logic puzzles to an undergraduate mathematics club. I was introduced by the logician Melvin Fitting (a former student of mine, who knows me extremely well). His introduction really captures the spirit of this book almost better than the book itself! He said, "I now introduce Professor Smullyan, who will prove to you that either he doesn't exist or you don't exist, but you won't know which."

239. Proof that Tweedledoo Exists _____

> (1) TWEEDLEDOO EXISTS
> (2) BOTH SENTENCES IN THIS BOX ARE FALSE

Let us first look at sentence (2). If it were true, then both sentences would be false, hence sentence (2) would be false, which is a contradiction. Therefore sentence (2) is false. Hence it is not the case that both sentences are false, so at least one of them is true. Since sentence (2) is not true, it must be sentence (1) which is true. Therefore Tweedledoo exists.

240. And What About Santa Claus? _____

There seems to be a lot of skepticism about the existence of Santa Claus. For example, in the Marx brothers movie *A Night at the Opera*, Groucho was going through a contract with Chico, and they came to one clause stating that if any of the parties participating in the contract is shown not to be in his right mind, the entire agreement is automatically nullified—this clause is known as the sanity clause. Chico says, 'You can't fool me—there ain't no Sanity Clause!"

I also recall in my high school days a joke going around about Mae West: Why can't Mae West be in the same telephone booth with Santa Claus? *Answer:* Because there is no Santa Claus. (This might aptly be called an "ontological" joke.)

Well, despite this modern skepticism, I will now give you three proofs which will establish beyond any reasonable possibility of doubt that Santa Claus does and must exist. These proofs are variants of a method, derived from J. Barkley Rosser, of proving anything whatsoever.

Proof One: We shall present this proof in the form of a dialogue.

First Logician / Santa Claus exists, if I am not mistaken.

Second Logician / Well of course Santa Claus exists, *if you are not mistaken.*

First Logician / Hence my statement is true.

Second Logician / Of course!

First Logician / So I was not mistaken—and you admitted that if I am not mistaken, then Santa Claus exists. Therefore Santa Claus exists.

Proof Two: The above proof is but a literary elaboration of the following proof of J. Barkley Rosser.

> IF THIS SENTENCE IS TRUE
> THEN SANTA CLAUS EXISTS

The idea behind this proof is the same as that of the proof that when an inhabitant of an island of knights and knaves says: "If I am a knight then so-and-so," then he must be a knight and the so-and-so must be true.

If the sentence is true, then surely Santa Claus exists (because if the sentence is true then it must also be true that if the sentence is true then Santa Claus exists, from which follows that Santa Claus exists), hence what the sentence says is the case, so the sentence is true. Hence the sentence is true and if the sentence is true then Santa Claus exists. From this it follows that Santa Claus exists.

Question / Suppose an inhabitant of an island of knights and knaves said, "If I'm a knight then Santa Claus exists." Would this prove that Santa Claus exists?

Answer / It certainly would. Since, however, Santa Claus doesn't exist, then neither a knight nor a knave could make such a statement.

Proof Three:

> THIS SENTENCE IS FALSE AND
> SANTA CLAUS DOES NOT EXIST

I leave the details to the reader.

Discussion. What is wrong with these proofs? Well, the underlying fallacy is exactly the same as in the reasoning of the suitor of Portia Nth: some of the sentences involved are not meaningful (see discussion in Chapter 15), hence should not be assumed to be either true or false.

The next proof we shall consider is based on a totally different principle.

241. Proof that Unicorns Exist. _____

I wish to prove to you that there exists a unicorn. To do this, it obviously suffices to prove the (possibly) stronger statement that there exists an *existing* unicorn. (By an existing unicorn I of course mean a unicorn which exists.) Surely if there exists an existing unicorn, then there must exist a unicorn. So all I have to do is prove that an existing unicorn exists. Well, there are exactly two possibilities:

(1) An existing unicorn exists.
(2) An existing unicorn does not exist.

Possibility (2) is clearly contradictory: How could an existing unicorn not exist? Just as it is true that a blue unicorn is necessarily blue, an existing unicorn must necessarily be existing.

Discussion. What is wrong with this proof? This proof is nothing more than the distilled essence of Descartes' famous ontological proof of the existence of God. Descartes defines God as a being which has all properties. Hence, by definition, God must also have the property of existence. Therefore God exists.

Immanuel Kant claimed Descartes' argument to be invalid on the grounds that existence is not a property. I believe there is a far more significant error in the proof. I shall not argue here the question of whether existence is or is not a property; the point I wish to make is that even if existence is a property, the proof is still no good.

Consider first my proof (sic) of the existence of a unicorn. As I see it, the real fallacy lies in the double meaning of the word "an," which in some contexts means "every" and in other contexts means "at least one." For example, if I say, "An owl has large eyes," what is meant is that owls have large eyes, or that all owls have large eyes, or that every owl has large eyes. But if I say, "An owl is in the house," I certainly do not mean that all owls are in this house, but only that there exists an owl who is in this house. So, when I say "an existing unicorn exists," it is not clear whether I mean that *all* existing unicorns exist or that there exists an existing unicorn. If I meant the first, then it is true—of course all existing unicorns exist; how could there be an existing unicorn who does not exist? But this does not mean that the statement is true in the second sense, that is, that there must exist an existing unicorn.

Similarly with Descartes' proof; all that properly follows is that all Gods exist, that is, that anything satisfying Descartes' definition of a God must also have the property of existence. But this does not mean that there necessarily exists a God.

242. Proof by Coercion. _____

There is the famous anecdote about Diderot paying a visit to the Russian Court at the invitation of the Empress. He made quite free with his views on atheism. The Empress herself was highly amused, but one of her councillors suggested that it might be desirable to put a check on these expositions of doctrine. They then conspired with the mathematician Euler, who was present at the occasion, and who himself was a believer. Euler announced that he had a proof of the existence of God which he would give before all the court, if Diderot desired to hear it. Diderot gladly consented. Well, Euler, taking advantage of Diderot's lack of knowledge of mathematics, advanced toward Diderot and said in a grave voice: "A squared minus B squared equals A minus B times A plus B—therefore God exists. Reply!" Diderot was embarrassed and disconcerted, while peals of laughter rose on all sides. He asked permission to return at once to France, and it was granted.

243. A Proof that You Are Either Inconsistent or Conceited. _____

I thought of this proof about thirty years ago and told it to several students and mathematicians. A few years ago someone told me that he had read it in some philosophical journal, but he could not recall the author. Anyway, here is the proof.

A human brain is but a finite machine, therefore there are only finitely many propositions which you believe. Let us label these propositions $p1$, $p2$, . . . , pn, where n is the number of propositions you believe. So you believe each of the propositions $p1$, $p2$, . . . , pn. Yet, unless you are conceited, you know that you sometimes make mistakes, hence not everything you believe is true. Therefore, if you are not conceited, you know that at least one of the propositions, $p1$, $p2$, . . . , pn is false. Yet you believe each of the propositions $p1$, $p2$, . . . , pn. This is a straight inconsistency.

Discussion. What is the fallacy of this argument? In my opinion, none. I really believe that a reasonably modest person has to be inconsistent.

B. MORE MONKEY TRICKS

244. Russell and the Pope. _____

One philosopher was shocked when Bertrand Russell told

him that a false proposition implies any proposition. He said, "You mean that from the statement that two plus two equals five it follows that you are the Pope?" Russell replied "Yes." The philosopher asked, "Can you prove this?" Russell replied, "Certainly," and contrived the following proof on the spot:

(1) Suppose $2 + 2 = 5$.
(2) Subtracting two from both sides of the equation we get $2 = 3$.
(3) Transposing, we get $3 = 2$.
(4) Subtracting one from both sides, we get $2 = 1$.

Now, the Pope and I are two. Since two equals one, then the Pope and I are one. Hence I am the Pope.

245. Which Is Better? _____

Which is better, eternal happiness or a ham sandwich? It would appear that eternal happiness is better, but this is really not so! After all, nothing is better than eternal happiness, and a ham sandwich is certainly better than nothing. Therefore a ham sandwich is better than eternal happiness.

246. Which Clock Is Better? _____

This one is due to Lewis Carroll. Which is better, a clock that loses a minute a day or a clock that doesn't go at all? According to Lewis Carroll the clock that doesn't go at all is better, because it is right twice a day, whereas the other clock is right only once in two years. "But," you might ask, "what's the good of it being right twice a day if you can't tell when the time comes?" Well, suppose the clock points to eight o'clock. Then when eight comes around, the clock is right. "But," you continue, "how does one know when eight o'clock does come?" The answer is very simple. Just keep your eye very carefully on the clock *and the very moment it is right* it will be eight o'clock.

247. Proof that There Exists a Horse with Thirteen Legs.

This proof is not original; it is part of the folklore of mathematicians.

We wish to prove that there exists at least one horse who has exactly thirteen legs. Well, paint all the horses in the universe either blue or red according to the following scheme: Before you paint the horse, count the number of its legs. If it has exactly thirteen legs, then paint it blue; if it has either fewer or more than thirteen legs, paint it red. You have now painted all horses in the universe; the blue ones have thirteen legs and the red ones don't. Well, pick a horse at random. If it is blue, then my assertion has been proven. If it is red, then pick a second horse at random. If the second horse is blue, then my assertion has been proven. But suppose the second horse is red? Ah, that would be a horse of a different color! But that's a contradiction, since the horse would be of the same color!

248.

I am reminded of a conundrum posed by Abraham Lincoln: If the tail of a dog was called a leg, how many legs would a dog have? Lincoln's answer was: "Four; calling the tail a leg doesn't mean that it is one."

249. My Favorite Method of All.

This is the best monkey trick I know. It is an absolutely unbeatable method of proving anything whatever. Its sole drawback is that only a magician can present it.

Here is what I do: Suppose I wish to prove to somebody that I am Dracula. I say, "The only logic you must know is that given any two propositions p and q, if p is true, then at least one of the two propositions p,q is true." Virtually everyone will assent to this. "Very well," I say, as I take a deck of cards out of my pocket, "as you can see, this card is red." I then place the red card face down on the left palm of the "victim" and I have him cover the back with his right hand. I continue: "Let p be the proposition that the card you are holding is red; let q be the proposition that I am Dracula. Since p is true, then you grant that either p or q is true?" He assents. "Well now," I continue, "p is obviously false—just turn over the card." He does so, and to his amazement the card is black! "Therefore," I conclude triumphantly, "q is true, so I am Dracula!"

C. SOME LOGICAL CURIOSITIES

In the last two sections we considered several invalid arguments which at first sight appeared to be valid. We shall now do the very opposite: we will consider some principles which at first seem downright crazy, but turn out to be valid after all.

250. The Drinking Principle. _____

There is a certain principle which plays an important role in modern logic and which some of my graduate students have affectionately dubbed "The Drinking Principle." Perhaps the reason it got its name is that I always preface the study of this principle with the following joke.

A man was at a bar. He suddenly slammed down his fist and said, "Gimme a drink, and give everyone elsch a drink, caush when I drink, everybody drinksh!" So drinks were happily passed around the house. Some time later, the man said, "Gimme another drink, and give everyone elsch another drink, caush when I take another drink, everyone takesch another drink!" So, second drinks were happily passed around the house. Soon after, the man slammed some money on the counter and said, "And when I pay, everybody paysh!"

This concludes the joke. The problem, now, is this: Does there really exist someone such that if he drinks, everybody drinks? The answer will surprise many of you.

A more dramatic version of this problem emerged in a conversation I had with the philosopher John Bacon: Prove that there is a woman on earth such that if she becomes sterile, the whole human race will die out.

A dual version of The Drinking Principle is this: Prove that there is at least one person such that if anybody drinks, then he does.

Solution. Yes, it really is true that there exists someone such that whenever he (or she) drinks, everybody drinks. It comes ultimately from the strange principle that a false proposition implies any proposition.

Let us look at it this way: Either it is true that everybody drinks or it isn't. Suppose it is true that everybody drinks. Then take any person—call him Jim. Since everybody drinks and Jim drinks, then it is true that if Jim drinks then everybody drinks. So there is at least one person—namely Jim—such that if he drinks then everybody drinks.

Suppose, however, that it is not true that everybody drinks; what then? Well, in that case there is at least one person—call him Jim—who doesn't drink. Since it is false that Jim drinks, then it is true that *if* Jim drinks, everybody drinks. So again there is a person—namely Jim—such that if he drinks, everybody drinks.

To summarize, call a person "mysterious" if he has the strange property that his drinking implies that everybody drinks. The upshot of the matter is that if everyone drinks, then anyone can serve as the mysterious person, and if it is not the case that everybody drinks, then any nondrinker can serve as the mysterious person.

As for the more dramatic version, by the same logic it follows that there is at least one woman such that if she becomes sterile, all women will become sterile (namely, any woman, if all women become sterile, and any woman who doesn't become sterile, if not all women become sterile). And, of course, if all women become sterile, the human race will die out.

As for the "dual" version, i.e., that there is someone such that if anybody at all drinks, then he does—either there is at least one person who drinks or there isn't. If there isn't, then take any person—call him Jim. Since it is false that someone drinks, then it is true that *if* someone drinks then Jim drinks. On the other hand, if there is someone who drinks, then take any person who drinks—call him Jim. Then it is true that someone drinks and it is true that Jim drinks, hence it is true that if someone drinks then Jim drinks.

Epilogue.

When I told The Drinking Principle to my students Linda Wetzel and Joseph Bevando, they were delighted. Shortly after, they wrote me a Christmas card in which they invented the following imaginary conversation (allegedly over dinner in the cafeteria).

Logician / I know a fellow who is such that whenever he drinks, everyone does.

Student / I just don't understand. Do you mean, everyone on earth?

Logician / Yes, naturally.

Student / That sounds crazy! You mean as soon as he drinks, at *just* that moment, *everyone* does?

Logician / Of course.

Student / But that implies that at some time, *everyone* was drinking at *once*. Surely that never happened!

Logician / You didn't listen to what I said.

Student / I certainly did—what's more, I have refuted your logic.

Logician / That's impossible. Logic cannot be refuted.

Student / Then how come I just did?

Logician / Didn't you tell me that you never drink?

Student / Uh . . . yes, I guess we'd better change the subject.

251. Is This Argument Valid? _____

I have seen many arguments in my life which seem valid but are really invalid. I only recently came across an argument which at first seems invalid (indeed, it seems like a joke) but turns out to be valid.

Incidentally, by a valid argument is meant one in which the conclusion necessarily follows from the premises; it is not necessary that the premises be true.

Here is the argument:[1]

(1) Everyone is afraid of Dracula.
(2) Dracula is afraid only of me.

Therefore I am Dracula.

Doesn't that argument sound like just a silly joke? Well it isn't; it is valid: Since everyone is afraid of Dracula, then Dracula is afraid of Dracula. So Dracula is afraid of Dracula, but also is afraid of no one but me. Therefore I must be Dracula!

So here is an argument which seems like a joke, but turns out not to be one—that's the funny part of it!

[1] I got it from the philosopher Richard Cartwright.

15. From Paradox to Truth

A. PARADOXES

252. The Protagoras Paradox.

Perhaps one of the earliest known paradoxes is about the Greek law teacher Protagoras, who took a poor but talented student and agreed to teach him without a fee on condition that after the student completed his studies and won his first law case, he would pay Protagoras a certain sum. The student agreed to do this. Well, the student completed his studies but did not take any law cases. Some time elapsed and Protagoras sued the student for the sum. Here are the arguments they gave in court.

Student / If I win the case, then by definition, I don't have to pay. If I lose the case, then I will not yet have won my first case, and I have not contracted to pay Protagoras until after I have won my first case. So whether I win the case or lose the case, I don't have to pay.

Protagoras / If he loses the case, then by definition he has to pay me (after all, this is what the case is about). If he wins the case, then he will have won his first case, hence he has to pay me. In either case, he has to pay me.

Who was right?

Discussion. I'm not sure I really know the answer to this dilemma. This puzzle (like the first puzzle of this book, concerning whether I was fooled or not) is a good prototype of a whole family of paradoxes. The best solution I ever got was from a lawyer to whom I posed the problem. He said: "The court should award the case to the student—the student shouldn't have to pay, since he hasn't yet won his first case. After the termination of the case, *then* the student owes money to Protagoras, so Protagoras should then turn around and sue the student a second time. This time, the court should award the case to Protagoras, since the student has now won his first case."

253. The Liar Paradox. _____

The so-called "Liar Paradox," or "Epimenides Paradox," is really the cornerstone of a whole family of paradoxes of the type known as "liar paradoxes." (Boy, that sounded pretty circular, didn't it?) Well, the original form of the paradox was about a certain Cretan named Epimenides, who said, "All Cretans are liars."

In this form, we really do not get a paradox at all—no more than we get a paradox from the assertion that an inhabitant of an island of knights and knaves makes the statement, "All people on this island are knaves." What properly follows is: (1) the speaker is a knave; (2) there is at least one knight on the island. Similarly, with the above version of the Epimenides paradox, all that follows is that Epimenides is a liar and that at least one Cretan is truthful. This is no paradox.

Now, if Epimenides were the *only* Cretan, then we would indeed have a paradox, just as we would have if a sole inhabitant of an island of knights and knaves said that all inhabitants of the island were knaves (which would be tantamount to saying that he is a knave, which is impossible).

A better version of the paradox is that of a person saying, "I am now lying." Is he lying or isn't he?

The following version is the version which we shall refer to as *the* liar paradox. Consider the statement in the following box:

> THIS SENTENCE IS FALSE

Is that sentence true or false? If it is false then it is true, and if it is true then it is false.

We shall discuss the resolution of this paradox a bit later.

254. A Double Version of the Liar Paradox. ___

The following version of the liar paradox was first proposed by the English mathematician P. E. B. Jourdain in 1913. It is sometimes referred to as "Jourdain's Card Paradox." We have a card on one side of which is written:

> (1) THE SENTENCE
> ON THE OTHER SIDE
> OF THIS CARD
> IS TRUE

Then you turn the card over, and on the other side is written:

> (2) THE SENTENCE
> ON THE OTHER SIDE
> OF THIS CARD
> IS FALSE

We get a paradox as follows: If the sentence is true, then the second sentence is true (because the first sentence says it is), hence the first sentence is false (because the second sentence says it is). If the first sentence is false then the second sentence is false hence the first sentence is not false but true. Thus the first sentence is true if and only if it is false, and this is impossible.

255. Another Version. _____

Another popular version of the liar paradox is given by the following three sentences written on a card.

```
(1) THIS SENTENCE CONTAINS FIVE WORDS
(2) THIS SENTENCE CONTAINS EIGHT WORDS
(3) EXACTLY ONE SENTENCE ON THIS CARD IS TRUE
```

Sentence (1) is clearly true, and sentence (2) is clearly false. The problem comes with sentence (3). If sentence (3) is true, then there are two true sentences—namely (3) and (1)—which is contrary to what sentence (3) says, hence sentence (3) would have to be false. On the other hand, if sentence (3) is false, then sentence (1) is the only true sentence, which means that sentence (3) must be true! Thus sentence (3) is true if and only if it is false.

Discussion. Now, what is wrong with the reasoning in these paradoxes? Well, the matter is subtle and somewhat controversial. There are those (philosophers, interestingly enough, rather than mathematicians) who rule out as legitimate any sentence which refers to itself. Frankly, I see this point of view as utter nonsense! In a self-referential sentence such as, "This sentence has five words," the meaning seems as clear and unequivocal as can be; just count the words and you will see the sentence must be true. Also, the sentence, "This sentence has six words," though false, is perfectly clear as to its meaning—it states that it has six words, which as a matter of fact it does not have. But there is no doubt about what the sentence says.

On the other hand, consider the following sentence:

```
THIS SENTENCE IS TRUE
```

Now, the above sentence does not give rise to any paradox; no logical contradiction results either from assuming the sentence to be true or from assuming the sentence to be false. Nevertheless, the sentence has no meaning whatsoever for the following reasons:

Our guiding principle is that to understand what it means for a sentence to be true, we must first understand the meaning of the sentence itself. For example, let X be the sentence: Two plus two equals four. Before I can understand what it means for X to be true, I must understand the meaning of every word which occurs in X, and I must know just what it is that X asserts. In this case, I do know the meaning of all the words in X, and I know that X means that two plus two equals four. And since I know that two plus two *does* equal four, then I know that X must be true. But I couldn't have known that X was true until I first knew that two plus two equals four. Indeed, I couldn't have even known what it *means* for X to be true unless I first knew what it means for two plus two to equal four. This illustrates what I mean when I say that the meaning of a sentence X being true is *dependent* on the meaning of X itself. If X should be of such a peculiar character that the very meaning of X *depends* on the meaning of X being true, then we have a genuinely circular deadlock.

Such is exactly the case with the sentence in the above box. Before I can know what it means for the sentence to be true, I must first understand the meaning of the sentence itself. But what *is* the meaning of the sentence itself; what does the sentence say? Merely that the sentence is true, and I don't yet know what it means for the sentence to be true. In short, I can't understand what it means for the sentence to be true (let alone whether it *is* true or not) until I first understand the meaning of the sentence, and I can't understand the meaning of the sentence until I first understand what it means for the sentence to be true. Therefore the sentence conveys no information whatsoever. Sentences having this feature are technically known as sentences which are not *well-grounded.*

The liar paradox (and all its variants) rest on the use of ungrounded sentences. (I am using "ungrounded" as short for "not well-grounded.") In number 253 the expression, "This sentence is false" is not well-grounded. In number 254, neither sentence on either side of the card is well-grounded. In number 255, the first two sentences are well-grounded, but the third sentence is not.

Incidentally, we can now say more as to how the suitor of Portia Nth got into trouble with his reasoning (see Chapter 5 on Portia's caskets). All the earlier Portias used only sentences which were well-grounded, but Portia Nth made skillful use of ungrounded sentences to bedazzle her suitor. The same fallacy occurs in the first few proofs of the last chapter.

256. But What About This One? _____

We return to our friends, Bellini and Cellini of the story of Portia's caskets. These two craftsmen made not only caskets, but also signs. As with the caskets, whenever Cellini made a sign, he inscribed a false statement on it, and whenever Bellini made a sign, he inscribed a true statement on it. Also, we shall assume that Cellini and Bellini were the only sign-makers of their time (their sons made only caskets, not signs).

You come across the following sign:

```
┌─────────────────────┐
│  THIS SIGN WAS      │
│  MADE BY CELLINI    │
└─────────────────────┘
```

Who made the sign? If Cellini made it, then he wrote a true sentence on it—which is impossible. If Bellini made it, then the sentence on it is false—which is again impossible. So who made it?

Now, you can't get out of this one by saying that the sentence on the sign is not well-grounded! It certainly is well-grounded; it states the historical fact that the sign was made by Cellini; if it was made by Cellini then the sign is true, and if it wasn't, the sign is false. So what is the solution?

The solution, of course, is that I gave you contradictory information. If you actually came across the above sign, then it would mean either that Cellini sometimes wrote true inscriptions on signs (contrary to what I told you) or that at least one other sign-maker sometimes wrote false statements on signs (again, contrary to what I told you). So this is not really a paradox, but a swindle.

Incidentally, have you yet figured out the name of this book?

257. Hanged or Drowned? _____

In this popular puzzle, a man has committed a crime punishable by death. He is to make a statement. If the statement is true, he is to be drowned; if the statement is false, he is to be hanged. What statement should he make to confound his executioners?

258. The Barber Paradox. _____

This is another well-known puzzle. It is given that a barber of a certain small town shaved all the inhabitants of the town who did not shave themselves, and never shaved any inhabitant who did shave himself. The question is whether the barber shaves himself or not. If he does, then he is violating the rule, since he is then shaving someone who shaves himself. If he doesn't, then he is again violating his rule, since he is failing to shave someone who is not shaving himself. So what should the barber do?

259. And What About This? _____

On an island of knights and knaves two inhabitants, A and B, make the following statements:

A: B is a knave.
B: A is a knight.

Would you say that A is a knight or a knave? What would you say about B?

SOLUTION TO PROBLEMS 257, 258, 259

257. _____

All he has to say is, "I will be hanged."

258. _____

The answer is that it is logically impossible that there exists any such barber.

259. _____

What you *should* say is that the author is lying again! The situation I described is quite impossible; it is really Jourdain's Double Card Paradox in a slightly different dress (see problem 254).

    If A is a knight then B is really a knave, hence A is not really a knight! If A is a knave, then B is not really a knave, he is a knight, hence his statement is true, which makes A a knight. Hence A cannot be either a knight or a knave without contradiction.

B. FROM PARADOX TO TRUTH

Someone once defined a paradox as a truth standing on its head. It is certainly the case that many a paradox contains an idea which with a little modification leads to an important new discovery. The next three puzzles afford a good illustration of this principle.

260. What Is Wrong with This Story? _____

Inspector Craig once visited a community and had a talk with one of the inhabitants, a sociologist named McSnurd. Professor McSnurd gave Craig the following sociological account:

"The inhabitants of this community have formed various clubs. An inhabitant may belong to more than one club. Each club is named after an inhabitant; no two different clubs are named after the same inhabitant, and every inhabitant has a club named after him. It is not necessary that a person be a member of the club named after him; if he is, then he is called *sociable*; if he isn't, then he is called *unsociable*. The interesting thing about this community is that the set of all unsociable inhabitants forms a club."

Inspector Craig thought about this for a moment, and suddenly realized that McSnurd couldn't have been a very good sociologist; his story simply didn't hold water. Why?

Solution. / This is really the Barber Paradox in a new dress.

Suppose McSnurd's story was true. Then the club of all unsociable inhabitants is named after someone—say Jack. Thus we will call this club "Jack's Club." Now, Jack is either sociable or unsociable, and either way we have a contradiction: Suppose Jack is sociable. Then Jack belongs to Jack's Club, but only unsociable people belong to Jack's Club, so this is not possible. On the other hand, if Jack is unsociable, then Jack belongs to the club of unsociable people, which means that Jack belongs to Jack's Club (which is the club of unsociable people), which makes Jack sociable. So either way we have a contradiction.

261. Is There a Spy in the Community? _____

Inspector Craig once visited a second community and spoke to an old friend of his, a sociologist named McSnuff. Craig and McSnuff had gone through Oxford together, and Craig knew him to be a man of impeccable judgment. McSnuff gave Craig the following account of this community:

"Like the other community, we have clubs, and each inhabitant has exactly one club named after him, and every club is named after someone. In this community, however, if a person is a member of a club, he can be so either secretly or openly. Anyone who is not openly a member of the club named after him is called *suspicious*. If anyone were known to secretly belong to the club named after him, he would be called a *spy*. Now, the curious thing about this community is that the set of all suspicious characters forms a club."

Inspector Craig thought about this for a moment, and realized that, unlike the last story, this story is perfectly consistent. Moreover, something interesting emerges from it—namely, that it is possible to deduce whether or not there are any actual spies in the community.

Are there?

Solution. The club of all suspicious characters is named after someone—call him John. Thus we will call this club "John's Club."

Now, either John himself is a member of John's Club or he isn't. Suppose he isn't. Then he can't be suspicious (because every suspicious person *is* a member of John's Club). This means that John is openly a member of John's Club. So if John is not a member of John's Club, then John is openly a member of John's Club, which is absurd. Therefore John must be a member of John's Club. Since every member of John's Club is suspicious, then John must be suspicious. Thus John is not openly a member of John's Club, yet he is a member, so he is secretly a member—in other words John is a spy!

We might remark that having solved the preceding problem, number 260, there is a simpler way of doing the immediate problem—namely to observe that if there were no spies in the community, then being suspicious would be no different than being unsociable, hence the set of all suspicious characters would be the same as the set of unsociable people, which would mean that the set of all *unsociable* people forms a club. But we proved in problem 260 that the set of all unsociable people cannot form a club. Therefore the assumption that there are no spies in the community leads to a contradiction, hence there must be a spy in the community (but in this proof, we have no idea who).

These two proofs afford a perfect illustration of what mathematicians mean by the terms "constructive proof" and "nonconstructive proof." The second proof is nonconstructive in the sense that although it showed that it couldn't be the case that there are no spies, it did not exhibit any actual spy. By contrast, the first proof is called constructive in that it actually exhibited a spy—namely the person (whom we called "John") after whom the club of suspicious characters is named.

262. Problem of the Universe. _____

There is a certain Universe in which *every* set of inhabitants forms a club. The Registrar of this Universe would like to name each club after an inhabitant in such a way that no two clubs are named after the same inhabitant and each inhabitant has a club named after him.

Now, if this Universe had only finitely many inhabitants, the scheme would be impossible (since there would be more clubs than inhabitants—for example, if there were just 5 inhabitants, there would be 32 clubs (including the empty set); if there were 6 inhabitants, there would be 64 clubs, and in general, if there are n inhabitants, there must be 2^n clubs). However, this particular Universe happens to contain *infinitely* many inhabitants, hence the Registrar sees no reason why his scheme should not be feasible. For trillions of years he has been trying to construct such a scheme, but so far every attempt has failed. Is the failure due to lack of ingenuity on the part of the Registrar, or is he attempting to do something inherently impossible?

Solution. He is attempting the impossible; this famous fact was discovered by the mathematician George Cantor. Suppose the Registrar could succeed in naming all the clubs after all the inhabitants in such a way that no two different clubs were named after the same inhabitant. Again, let us call an inhabitant *unsociable* if he is not a member of the club named after him. The collection of all unsociable inhabitants of this Universe certainly constitutes a well-defined set, and we are given that *every* set of inhabitants forms a club. Therefore we have the impossible club of all unsociable inhabitants—impossible for the same reason as that of problem 260 (this club must be named after somebody, and this somebody cannot be either sociable or unsociable without entailing a contradiction).

263. Problem of the Listed Sets. _____

Here is the same problem in a different dress; some of the notions involved will pop up again in the next chapter.

A certain mathematician keeps a book called *The Book of Sets*. On each page is written a description of a set of numbers. We use the word "numbers" to mean the positive whole numbers $1, 2, 3, \ldots n, \ldots$. Any set which is listed on any page is called a *listed* set. The pages are numbered consecutively.

The problem is to describe a set which is not listed on any page of the book.

Solution. Given any number n, call n an *extraordinary* number if n belongs to the set listed on page n; call n an *ordinary* number if n does not belong to the set listed on page n.

The set of ordinary numbers cannot possibly be listed; if it were, the number of the page on which it was listed couldn't be either ordinary or extraordinary without entailing a contradiction.

16. Gödel's Discovery

A. GÖDELIAN ISLANDS

The puzzles of this section are adaptions of a famous principle discovered by the mathematical logician Kurt Gödel which we discuss at the end of the chapter.

264. The Island G.

A certain island G is inhabited exclusively by knights who always tell the truth and knaves who always lie. In addition, some of the knights are called "established knights" (these are knights who in a certain sense have "proved themselves") and certain knaves are called "established knaves." Now, the inhabitants of this island have formed various clubs. It is possible that an inhabitant may belong to more than one club. Given any inhabitant X and any club C, either X claims that he is a member of C or he claims that he is not a member of C.

We are given that the following four conditions, E_1, E_2, C, G, hold.

E_1: The set of all established knights forms a club.
E_2: The set of all established knaves forms a club.
C (*The Complementation Condition*): Given any club

C, the set of all inhabitants of the island who are not members of C form a club of their own. (This club is called the *complement* of C and is denoted by C̄.)

    G (*The Gödelian Condition*): Given any club C, there is at least one inhabitant of the island who claims that he is a member of C. (Of course his claim might be false: he could be a knave.)

264a. _____

(After Gödel) (i) Prove that there is at least one unestablished knight on the island.

    (ii) prove that there is at least one unestablished knave on the island.

264b. _____

(After Tarski) (i) Does the set of all knaves on the island form a club?

    (ii) Does the set of all knights on the island form a club?

Solution to 264a. By condition E_1, the set E of all established knights forms a club. Hence by condition C, the set Ē of all people on the island who are *not* established knights also forms a club. Then by condition G, there is at least one person on the island who claims to be a member of the club Ē—in other words, he claims that he is *not* an established knight.

    Now, a knave couldn't possibly claim that he is not an established knight (because it is true that a knave is not an established knight), hence the speaker must be a knight. Since he is a knight, then what he says is true, so he is not an established knight. Therefore the speaker is a knight but not an established knight.

    By condition E_2, the set of established knaves forms a club. Therefore (by condition G) there is at least one person on the island who claims to be an established knave (he claims to be a member of the club of established knaves). This person cannot be a knight (since no knight would claim to be any kind of a knave) hence he is a knave. Therefore his statement is false, so he is not an established knave. This means that he is a knave but not an established knave.

Solution to 264 b. If the set of knaves formed a club, then at least one inhabitant would claim to be a knave, which neither a knight nor a knave could do. Therefore the set of knaves does not form a club.

If the set of knights formed a club, then the set of knaves also would (by condition C), hence the knights don't form a club either.

Remarks. (1) Problem 264b affords an alternative solution to problem 264a, which, though nonconstructive, may be somewhat simpler.

If every knight were established, then the set of knights would be the same as the set of established knights, but this is impossible because the set of established knights forms a club (by condition E_1) but the set of knights doesn't (by problem 264b). Thus the assumption that all knights are established leads to a contradiction, hence there must be at least one unestablished knight. Similarly, if every knave were established, then the set of established knaves would be the same as the set of knaves, which cannot be, since the set of established knaves forms a club whereas the set of knaves doesn't.

By contrast with this proof, our first proof tells us specifically that anyone who claims that he is not an established knight must be an unestablished knight, and anyone who claims to be an established knave must be an unestablished knave.

(2) Our proof that the set of knaves does not form a club used only condition G; conditions E_1, E_2, and C were not needed for this. Thus condition G alone implies that the knaves don't form a club. Actually, condition G is *equivalent*

to the statement that the knaves don't form a club, for suppose we are given that the set of knaves doesn't form a club; we can derive condition G as follows:

Take any club C. Since the set of knaves is not a club, then C is not the set of all knaves. Hence either some knight is in C or some knave is outside C. If some knight is in C, he would certainly claim to be in C (since he is truthful). If some knave were outside C, he would also claim to be in C (since he lies). So in either case, *someone* claims to be in C.

265. Gödelian Islands in General. _____

Consider now an arbitrary knight-knave island with clubs. (By a knight-knave island, we mean, of course, an island inhabited exclusively by knights and knaves.) We shall call the island a *Gödelian* island if condition G holds, i.e., for every club C, there is at least one inhabitant who claims to be a member of the club.

Inspector Craig once visited a knight-knave island which had clubs. Craig (who, incidentally, is a highly cultured gentleman whose theoretical interests are as strong as his practical ones) was curious to know whether or not he was on a Gödelian island. He found out the following information.

Each club is named after an inhabitant and each inhabitant has a club named after him. An inhabitant is not necessarily a member of the club named after him; if he is, he is called *sociable,* if he isn't he is called *unsociable.* An inhabitant X is called a *friend* of an inhabitant Y if X testifies that Y is sociable.

Craig still did not know whether or not he was on a Gödelian island until he found out that the island satisfied the following condition, which we will call *condition H.*

H: For any club C, there is another club D such that every member of D has at least one friend in C, and every nonmember of D has at least one friend who is not a member of C.

From this condition H, Craig could deduce whether this island was Gödelian.

Is it?

Solution. Yes, it is. Take any club C. Let D be a club given by condition H. This club D is named after someone—say John. Either John belongs to club D or he doesn't.

    Suppose he does. Then he has a friend—call him Jack—in club C who testifies that John is sociable. Since John does belong to D, then John really is sociable, hence Jack is a knight. So Jack is a knight who belongs to club C, so Jack will claim he belongs to club C.

    Suppose John doesn't belong to club D. Then John has a friend—call him Jim—who is not a member of C, and Jim claims that John is sociable. Since John is not a member of club D, then John is actually unsociable, hence Jim is a knave. So Jim is a knave who is not in club C, hence Jim would lie and claim that he is in club C. So whether John belongs to club D or doesn't belong to club D, there is an inhabitant who claims to be a member of club C.

    Remarks. Combining the results of 264 and 265, we see that given any island satisfying conditions E_1, E_2, C, and H, there must be both an unestablished knight and an unestablished knave on the island. This result is really a disguised form of Gödel's famous incompleteness theorem, which we will consider again in Section C of this chapter.

    Incidentally, if you would like to try a *really* tough problem on one of your friends, just give him an island with conditions E_1, E_2, C, and H (don't mention G), and pose problem 264. It would be interesting to see if he comes up with condition G himself.

B. DOUBLY GÖDELIAN ISLANDS

The puzzles of this section are of more specialized interest and might best be postponed until after section C.

By a "doubly Gödelian island" we shall mean a knight-knave island with clubs such that the following condition GG is satisfied:

GG: Given any two clubs C_1, C_2, there are inhabitants A,B such that A claims that B is a member of C_1 and B claims that A is a member of C_2.

As far as I know, condition GG does not imply condition G, nor does condition G imply condition GG; they appear to be quite independent. Thus (as far as I know) a doubly Gödelian island is not necessarily a Gödelian island.

The subject of doubly Gödelian islands is a pet hobby of mine. The puzzles involved bear the same sort of relation to the Jourdain Double Card Paradox (see problem 254 of the preceding chapter) as the puzzle of Gödelian islands bears to the liar paradox.

266. The Doubly Gödelian Island S. _____

I once had the good fortune to discover a doubly Gödelian island S in which conditions E_1, E_2, and C of island G all held.

  (a) Can it be determined whether there is an unestablished knight on S? What about an unestablished knave?
  (b) Can it be determined whether the knights of island S form a club? What about the set of knaves?

Solution. Let us first consider part (b). If the set of knights forms a club, then so does the set of knaves (by condition C), and if the set of knaves forms a club, so does the set of knights (again by conditions C). So if either of these two sets formed a club, they both would. Well, suppose they both do. Then by condition GG there must be inhabitants A,B who make the following claims:

A: B is a knave.
B: A is a knight.

This is an impossible situation, as we showed in the solution of problem 259 in the last chapter. The conclusion, therefore, is that neither the set of knights nor the set of knaves can form a club.

As for part (a) we can now solve it by either of two methods; the first is simpler, our having solved part (b), but the second is more instructive.

Method One: Since the set of knights does not form a club and the set of established knights does, then the two sets are different, hence not all the knights are established. Similarly with "knaves."

Method Two: Since the set of established knights forms a club, so does the set of all the inhabitants who are not established knights. Taking these two clubs for C_1, C_2, we have (by condition GG) inhabitants A,B who make the following claims:

A: B is an established knight.
B: A is not an established knight.

We leave it to the reader to verify that at least one of the two speakers A,B must be an unestablished knight (more specifically, if A is a knight then he is not an established knight, and if A is a knave then B must be an unestablished knight). The interesting thing is that although we know that one of A,B is an unestablished knight, we have no idea which one. (The situation is exactly like that of problem 134, the double casket problem of Bellini and Cellini; one of the caskets must be a Bellini, but there is no way to tell which.) Similarly, since the established knaves form a club, so does the set of all inhabitants who are not established knaves. Therefore (again by GG) there must be two speakers A,B who say:

A: B is an established knave.
B: A is not an established knave.

From this it follows that if B is a knave then he is an unestablished knave, and if B is a knight then A is an unestablished knave (again, we leave the proof of this to the reader), so in either case, either A or B is an unestablished knave, but we don't know which. (This problem is really the same as the double casket problem 135 of Bellini and Cellini.)

267. The Island S¹. _____

I once discovered another doubly Gödelian island S^1 which intrigued me even more. Conditions E_1, E_2 both hold for this island, but it is not known whether condition C holds or not. (We recall that condition C is that for any club C, the set of people not in C forms a club).

It appears impossible to prove that there is an unestablished knight on island S^1, or to prove that there is an unestablished knave. It also appears impossible to prove that the knights don't form a club, or to prove that the knaves don't form a club. However, the following *can* be proved:

(a) Prove that either there is an unestablished knight or an unestablished knave on this island.
(b) Prove that it is impossible that both the knights form a club and also the knaves form a club.

Solution. We will first do (b). Suppose that the knights formed a club *and* the knaves formed a club. Then there would be inhabitants A,B such that A claims B is a knave and B claims A is a knight, which we know to be impossible (see preceding.problem, or problem 259 of the last chapter). Thus it cannot be that the knights form a club and also that the knaves form a club; either the knights don't form a club or the knaves don't form a club. If the knights don't form a club, then there must be an unestablished knight (since the established knights *do* form a club); if the knaves don't form a club, then there must be an unestablished knave. But we can't tell which. This then also proves (a).

An alternative (and more interesting) method of proving that there is either an unestablished knight or an unestablished knave is this:

Since the established knights form a club and the established knaves form a club, then there are inhabitants A,B who say:

A: B is an established knave.
B: A is an established knight.

Suppose A is a knight. Then his statement is true, hence B is an established knave, so B's statement is false, hence A is not an established knight. So in this case, A is an unestablished knight. If A is a knave, then B's statement is false, so B is a knave. Also A's statement is false, so B is not an established knave. So, in this case, B is an unestablished knave.

Therefore either A is an unestablished knight or B is an unestablished knave (but again, we don't know which).

This problem again is like one of the double casket problems (number 136 of Chapter 9), in which one of the two caskets (we don't know which) was made by either Bellini or Cellini (but again we don't know which).

268. Some Unsolved Problems. _____

I have thought of a few problems concerning Gödelian and doubly Gödelian islands which I have not tried to solve; I feel it might be fun for the reader to try his hand at some original work.

268a. _____

I have stated that *as far as I know,* neither of the conditions G,GG imply the other. Can you prove that my conjecture is correct? (Or maybe disprove it, but I think that highly unlikely.) To do this you must construct an island in which G holds but GG does not, and construct an island in which

GG holds but G does not. By *constructing* an island I mean specifying all the inhabitants, then specifying which ones are knights and which ones are knaves and which sets of people form clubs and which ones do not. (Which knights and knaves are established has no bearing on this problem.)

268b. _____

Can you prove (or disprove) my conjecture that on island S^1 there needn't be an unestablished knight and there needn't be an unestablished knave (though, of course, there must be one or the other)? That is, can you construct an island satisfying E_1, E_2, and GG in which there are knights but no unestablished ones? Can you construct one in which there are knaves but no unestablished ones? (This time, in constructing such islands, you must specify not only the knights, knaves, and clubs, but also which knights and knaves are established.)

268c. _____

Assuming all these islands can be constructed (which I am morally certain is the case, even though I have not verified it), in each case what is the minimum number of inhabitants the island must have? Can you prove in each case that no smaller number will work?

C. GÖDEL'S THEOREM

269. Is This System Complete? _____

A certain logician keeps a book called *The Book of Sentences*. The pages of the book are numbered consecutively, and each page has exactly one sentence written on it. No sentence appears on more than one page. Given any sentence X, the number of the page on which it is written is called the *page number* of X.

Every sentence of the book, of course, is either true or false. Some of the true sentences are quite self-evident to this logician, and he has taken these self-evident truths as axioms of his logic system. This system also contains certain rules of reasoning which enable him to *prove* various true sentences from the axioms and to *disprove* various false ones. The logician is quite confident that his system is *correct* in the sense that every sentence which is provable in the system is indeed a true sentence, and every sentence which is disprovable in the system is a false one, but he is uncertain whether his system is *complete* in the sense that all the true sentences are provable and all the false are disprovable. Are all the true sentences provable in his system? Are all the false sentences disprovable in the system? These are the questions the logician would like to have answered.

Well, the logician also has a second book called *The Book of Sets*. This book also has all its pages consecutively numbered, and each page contains a description of a set of numbers. (We here use the word "numbers" to mean the positive whole numbers $1, 2, 3, \ldots, n, \ldots$.) Any set of numbers which is described anywhere in this book we will call a *listed* set.

Given any number n, it may happen that the set listed on page n (of *The Book of Sets*) contains n itself as a member; if this happens we will call n an *extraordinary number*. Also, given any numbers n, h, we will call h an *associate* of n if the sentence on page h (of *The Book of Sentences*) asserts that n is extraordinary.

We are given that the following four conditions hold:

E_1: The set of page numbers of all provable sentences is a listed set.

E_2: The set of page numbers of all the disprovable sentences is a listed set.

C: For any listed set A, the set $\bar{A}$ of all numbers not in A is a listed set.

H: Given any listed set A, there is another listed set B such that every number in B has an associate in A and every number outside B has an associate outside A.

These four conditions are sufficient to answer the logician's questions: Is every true sentence provable in the system? Is every false sentence disprovable in the system? It also can be determined whether or not the set of page numbers of all the true sentences is a listed set and whether the set of page numbers of all the false sentences is a listed set.

How can this be done?

Solution. This is nothing more than the Gödelian island puzzles of Section A in a different dress. In our present setup, the page numbers of the true sentences play the role of the knights; those of the false sentences, the knaves; those of the provable sentences, the established knights; and those of the disprovable sentences, the established knaves. The listed sets play the role of the clubs. The notion of a set being listed on a page bearing a given number plays the role of a club being named after a given inhabitant; hence the extraordinary numbers play the role of the sociable people, and the notion of "associate" plays the role of "friend."

The first thing we must do to solve the present problem is to prove the analog of condition G, which is this:

Condition G: For any listed set *A*, there is a sentence which is true if and only if its own page number lies in *A*.

To prove condition G, take any listed set *A*. Let *B* be a set given by condition H; let *n* be the number of a page on which *B* is listed. By condition H, if *n* lies in *B*, then *n* has an associate *h* in *A*; if *n* lies outside *B*, then *n* has an associate *h* outside *A*. We assert that the sentence *X* on page *h* is the sentence we seek.

The sentence *X* says that *n* is extraordinary—in other

words that n does lie in B (since B is the set listed on page n). If X is true then n really does lie in B, hence h lies in A. So if X is true, then its page number h does lie in A. Suppose X is false. Then n does not lie in B, hence h lies outside A. Thus X is true if and only if its page number lies in A.

Condition G having been proved, the logician's questions are now easily answered: We are given that the set A of page numbers of all the provable sentences is a listed set, hence by condition C, so is the set $\bar{A}$ of all numbers which are not page numbers of provable sentences, therefore, (by condition G) there is a sentence X which is true if and only if the page number of X belongs to $\bar{A}$. Now, to say that the page number of X belongs to $\bar{A}$ is to say that the page number of X doesn't belong to A, which is to say that X is not provable (since A consists of the page numbers of those sentences which *are* provable). Thus X is true if and only if X is not provable. This means that either X is true and not provable or X is false but provable. We are given that no false sentence is provable in the system, hence X must be true but not provable in the system.

As for obtaining a false sentence which is not disprovable, we now take A to be the set of page numbers of all the sentences which are disprovable. Applying condition G, we get a sentence Y which is true if and only if its page number is the page number of a disprovable sentence—in other words, Y is true if and only if Y is disprovable. This means that Y is either true and disprovable or false and not disprovable. The first possibility is out, since no disprovable sentence is true, hence Y must be false but not disprovable in the system.

As to the other questions, if the set of page numbers of all the false sentences were a listed set, then there would be a sentence Z which is true if and only if its page number is the page number of a false sentence—in other words, Z would be true if and only if Z is false, and this is impossible. (It would be like the sentence: "This sentence is false.")

Therefore the set of page numbers of all the false sentences is not a listed set. Then by condition C, the set of page numbers of the true sentences is not a listed set either.

270. Gödel's Theorem. _____

The above puzzle is really a form of Gödel's famous Incompleteness Theorem.

In 1931 Kurt Gödel came out with the startling discovery that in a certain sense, mathematical truth cannot be completely formalized. He showed that for a wide variety of mathematical systems—systems meeting certain very reasonable conditions—there must always be sentences which, though true, cannot be proved from the axioms of the system! Thus no formal axiom system, no matter how ingeniously constructed, is adequate to prove all mathematical truths. Gödel first proved this result for the celebrated system *Principia Mathematica* of Whitehead and Russell, but, as I said, the proof goes through for many different systems. In all these systems, there is a well-defined set of expressions called *sentences* and a classification of all sentences into *true* sentences and *false* sentences. Certain true sentences are taken as axioms of the system, and precise rules of inference are given enabling one to prove certain sentences and disprove others. In addition to sentences, the system contains names of various sets of (positive, whole) numbers. Any set of numbers which has a name in the system we might call a *nameable* or *definable* set of the system (these are the sets which we call the "listed" sets in the above puzzle). Now, the point is that it is possible to number all the sentences and to list all the definable sets in an order such that the conditions E_1, E_2, C, and H of our puzzle hold. (The number assigned to each sentence, which we called the "page number," is technically called the *Gödel number* of the sentence.) To establish conditions C and H is really a very simple matter, but to establish conditions E_1 and E_2 is quite a lengthy affair, though elementary in principle.[1] Anyway, once these four conditions are established, they lead to the construction of a sentence which is true but not provable in the system.

The sentence X in question might be thought of as asserting its own unprovability; such a sentence must in fact be true but not provable (just as a person on island G who asserts that he is not an established knight must, in fact, be a knight, but not an established one).

One might ask the following question: Since Gödel's sentence X (which asserts its own unprovability) is known to be true, why not add it as a further axiom to the system? Well, one can of course do this, but then the resulting enlarged system also satisfies conditions E_1, E_2, C and H, hence one can obtain another sentence X^1 which is both true but unprovable in the enlarged system. Thus, in the enlarged system, one can prove more true sentences than in the old system, but still not all true sentences.

I might remark that my account of Gödel's method departs somewhat from Gödel's original one—primarily in that it employs the notion of *truth*, which Gödel did not do. Indeed, Gödel's theorem in its original form did not say that there was a sentence which is true but not provable, but rather that under a certain reasonable assumption about the system, there must be a sentence (which Gödel actually exhibited) which is neither provable nor disprovable in the system.

A strict formalization of the notion of *truth* was done by the logician Alfred Tarski, and it was he who showed that for these systems, the set of Gödel numbers of the true sentences is not definable in the system. This is sometimes paraphrased: "For systems of sufficient strength, truth of

[1]Concerning condition H, for each number n, there is the sentence which asserts that n is extraordinary; this sentence (like every other sentence) has a Gödel number—call this number n*. Well, it turns out that for any definable set A, the set of all numbers n such that n* is in A—this set B is also definable. Since n* is an associate of n, condition H is fulfilled.

the sentences of the system is not definable within the system."

271. Last Words. _____

Consider the following paradox:

> THIS SENTENCE CAN NEVER BE PROVED

The paradox is this: If the sentence is false, then it is false that it can never be proved, hence it *can* be proved, which means it must be true. So, if it is false, we have a contradiction, therefore it must be true.

Now, I have just proved that the sentence is true. Since the sentence is true, then what it says is really the case, which means that it can never be proved. So how come I have just proved it?

What is the fallacy in the above reasoning? The fallacy is that the notion of *provable* is not well defined. One important purpose of the field known as "Mathematical Logic" is to make the notion of *proof* a precise one. However, there has not yet been given a fully rigorous notion of *proof* in any absolute sense; one speaks rather of provability *within a given system.* Now suppose we have a system —call it system S—in which the notion of *provability within the system S* is clearly defined. Suppose also that the system S is correct in the sense that everything provable in the system is really true. Now consider the following sentence:

> THIS SENTENCE IS NOT PROVABLE IN SYSTEM S

We now don't have any paradox at all, but rather an interesting truth. The interesting truth is that the above sentence must be a true sentence which is not provable in system S. It is, in fact, a crude formulation of Gödel's sentence X, which can be looked at as asserting its own unprovability, not in an absolute sense, but only within the given system.

I might also say just a little about the "doubly Gödelian" condition which I analyzed in Section B. The fact is that the various systems for which Gödel's result goes through are not only Gödelian" in the sense that given any definable set A there is a sentence which is true if and only if its Gödel number is in A, but these systems are what I might call "doubly Gödelian," by which I mean that given any two definable sets A,B, there are sentences X,Y such that X is true if and only if the Gödel number of Y is in A, and such that Y is true if and only if the Gödel number of X is in B. From this (using conditions E_1, E_2, and C) one can construct a pair X,Y such that X asserts that Y is provable (by which I mean that X is true if and only if Y is provable) and Y asserts that X is not provable; one of them (we don't know which) must be true but not provable. Or we can construct a pair X,Y such that X asserts that Y is disprovable and Y asserts that X is not disprovable—from which follows that at least one of them (we don't know which) must be false but not disprovable. Or again, (even without using condition C) we can construct a pair X,Y such that X asserts that Y is provable and Y asserts that X is disprovable; one of them (we don't know which) is either true but not provable, or false but not disprovable (but again we don't know which).

Oh, one last thing, before I forget: What is the name of this book? Well, the name of this book is: "What Is the Name of This Book?"